ELEMENTARY SCHOOL LESSON PLANS KINDERGARTEN TO SECOND GRADE

Integrating the Learning of MATHEMATICS with the Game of BASEBALL

Incorporating Florida's State Standards and Benchmarks for Mathematics as a Foundation for Instruction and Learning

by
James Schoedler, PhD, Physical Education
Manatee County, Bradenton, Florida

VOLUME I

ABOUT THE AUTHOR

Jim Schoedler received a B.S. from West Chester University and a Masters and PhD from the University of Maryland in Physical Education. He has focused his teaching career in physical education at the elementary level both in Delaware and Florida. He has had administrative experience at the school district level in curriculum and staff development. He has written numerous articles in professional journals and published with Collier-MacMillan, *Physical Skills for Young Children.*

He was a member of professional associations and has presented at numerous state, regional and national conventions in physical education and learning disabilities. He currently teaches elementary physical education in Manatee County, Bradenton, Florida.

DEVELOPMENT AND REVIEW PROCESS

The Baseball/Math curriculum presents a unique approach of integrating a sport with the teaching of mathematics. The lesson plans were reviewed by classroom teachers during their development with all suggestions being incorporated. The lesson plans were then pilot-taught by teachers at the K, 1, and 2 grade levels with all suggested improvements and changes included in revisions. The entire curriculum was reviewed by a county mathematics coordinator and approved for content relevancy and authenticity linking the state standards and benchmarks to the baseball-related activities.

UNIQUE FEATURES

- Newly developed Big Ideas and Benchmarks from the state of Florida for mathematics, K-2, are presented on each lesson plan page.
- A consistent lesson plan outline that is easy to use is offered and codes the Big Idea/Benchmarks for convenient recordkeeping.
- Classroom exercises are presented and related to physical qualities that baseball players need to perform well: flexibility and strength.
- An extensive appendix is provided and coded to specific lesson plans, including websites, references, terminology, etc.
- A pre-discussion to the lesson and follow-up variations are also included with each plan, which allows for teaching the lesson concept numerous times.
- Opportunities for all subject areas to achieve interdisciplinary methods are provided through the sample activities.
- Cognitive disabled learners will be extremely motivated to learn mathematic concepts through a sport theme and sensory-motor involvement in the activities.

D E D I C A T I O N

To my wife Nancy, for our fulfilling life together in work, play and family.

For her support, ideas and patience during the development of this document.

To our sons Scott and Craig and their families.

To Rob Gorley and Ronna Moore without whom this document would not have become a reality.

TABLE OF CONTENTS

ACKNOWLEDGEMENTS

The author wishes to recognize the following professionals for their contributions to this document.

- Critical review of lesson plans:
  Nikki Cucci Randy Stowers
  Laura Lawson Nancy Schoedler

- Piloting of lesson plans:
  Cari Losada, Kindergarten
  Vicki Paoli, Kindergarten
  Nikki Cucci, 1st grade
  Nancy Schoedler, 1st grade
  Randy Stowers, 2nd grade
  Jeff Gilmore, 2nd grade

- Computer Typing: Pam Keller

- Computer Graphics: Robert Fleming

- Editing: Agatha Tresky

- Math Curriculum Specialist: Joe McNaughton

- Tampa Bay Rays:
  Director of Community Relations—Suzanne Murchland
  Community Relations—Barry Jones

Published and Distributed By:
Baseball Curriculums, Inc.
P.O. Box 725
Tallevast, FL 34270
E-mail: baseballcurriculums@gmail.com

INTRODUCTION

Learning takes place when concepts are related to a personal or motivating topic such as the game of baseball. This curriculum guide has been written **by** teachers **for** teachers. It has been written for elementary school classroom teachers, ESE teachers, and physical education teachers. Anyone dedicated to the teaching and learning of elementary age children could benefit from this unique concept of presenting academics through the socially motivating game of baseball. Boys and girls of all ages will inherently be more enthusiastic about learning if it is related to a sport.

This curriculum guide will ensure that the state curriculum standards and benchmarks for each academic area will be taught since the lesson plans and baseball information presented are directly related. The activities and information presented in this guide are mere examples for the classroom teachers to support and supplement their own instructional methodology in achieving the state curriculum standards and benchmarks. Successful curriculum strategies are those whose efforts are coordinated toward student achievement and have state learning standards and guidelines as a foundation.

PREFACE

This curriculum guide contains lesson plans for the state curriculum standards and benchmarks in Mathematics. The unique aspect is that each lesson is taught with some aspect of the game of baseball as an underlying theme. The lessons are organized by grades Kindergarten through 2nd, which parallel the current and latest state curriculum standards and benchmarks.

The lessons are designed for teachers of all experience levels—classroom, special education, physical education, etc. The lessons can also be a valuable resource for elementary principals and supervisors who encourage the integration of academics with special area subjects.

SUPPORTING STATEMENTS

Piloting Teacher

"Great ideas! The students were excited to do math activities because of all the baseball "stuff". It was something that most could relate to, having played baseball themselves or having watched games. These lessons are great supplemental activities for teaching grade level math concepts."

Nikki Cucci
1st Grade Teacher
Freedom Elementary School
Manatee County

Piloting Teacher

"Baseball math is a wonderful collection of lessons that would be useful throughout the school year. The varied lessons and activities correlate with the benchmarks at different parts of the school year. Students love the thematic baseball approach to math and are totally engaged during lessons and activities. Reluctant students that enjoy baseball are interested and work harder on these lessons than standard math lessons."

Randy Stowers
2nd Grade Teacher
Freedom Elementary School
Manatee County

Piloting Teacher Comments About Students Reaction To Lesson Plan Activities

- "Children love pretending to buy things and worked hard at understanding."
- "Great activity for understanding money and making change."
- "Good activity to explain a fraction."
- "Students enjoyed and understood concept of fraction equaling parts of a whole."
- "Great activity to make students visualize addition and comparing similar amounts."

- “Good lesson for the beginning of the year.”
- “Students cheered when teacher announced, ‘now we are going to do baseball math.’”
- “Student said, ‘now I’ll be able to keep score by myself when I go to a baseball game with my Dad.’”

Principal

“I reviewed your curriculum for the teaching of mathematics and congratulate you on a job well done. We are happy to pilot your curriculum with one of our kindergarten classes and expect great results. Your curriculum captures not only the interest of the students, but their desire to learn about our all-American past-time of baseball. I especially like the way you sequenced all of the skills, building on the previous lesson’s objectives. Good luck with this creative product.

Ronna Moore
Principal, McNeal Elementary
Manatee County

University Professor

“It is wonderful to see math incorporated into activities that children enjoy. Standards-based math activities synthesized with the joy of play make this curriculum supplement a winner for both students and teachers. The curriculum will potentially enable a variety of young students to experience math meaningfully in a context beyond their school classroom. The baseball theme may also provide a bridge to involve other family members in students’ learning of standards-based math with the added benefit of physical activity. Dr. Schoedler has developed an extensive set of plans that will be useful to teachers in a variety of contexts. The organization and creativity of the lessons will have great appeal to those educators who are looking for alternative ways to reach their students with systematic standards-based math instruction.

Tary Wallace, Ph.D.
Assistant Professor
Measurement and Research, College of Education
University of South Florida

Former Florida Assistant Commissioner of Education,
Retired CEO, University of South Florida, Sarasota—Manatee Campus

"We know that children learn best when they can relate to the examples and information used to convey the concepts. The popularity of baseball and softball among girls and boys as well as across socio-economic classes means many children know about the game and have played it. Dr. Schoedler connected this American pastime with the math concepts within the game and it works. We know that increased math proficiency is a cornerstone to emerging professions as future climate monitors, space explorers, or medical pioneers. I applaud Dr. Schoedler's ingenuity in creating a pathway to meet the state's mathematics standards for young children."

Dr. Laurey T. Stryker

PURPOSE

There are many purposes of this curriculum guide for administrators, teachers, students, and parents, some of which include:

- To provide classroom teachers with a structured approach to following the state curriculum standards and benchmarks in their daily instructional plans.
- To provide teachers with specific activities related to the game of baseball in the teaching of mathematics.
- To offer students a fun, motivating way of learning academics.
- To relate students previous or wishful experiences in playing baseball or softball to learning academics.
- To provide principals and county administrators the assurance that the state curriculum standards and benchmarks will be reviewed and implemented in the teaching of mathematics.
- To stimulate unmotivated students with the motivation needed to become engaged and excited about learning.
- To offer special area teachers—Art, Music, Physical Education, Computer, Media, etc.—with the stimulus to become involved with the process of integrating their subject area with academics.

BASEBALL CURRICULUM MODEL

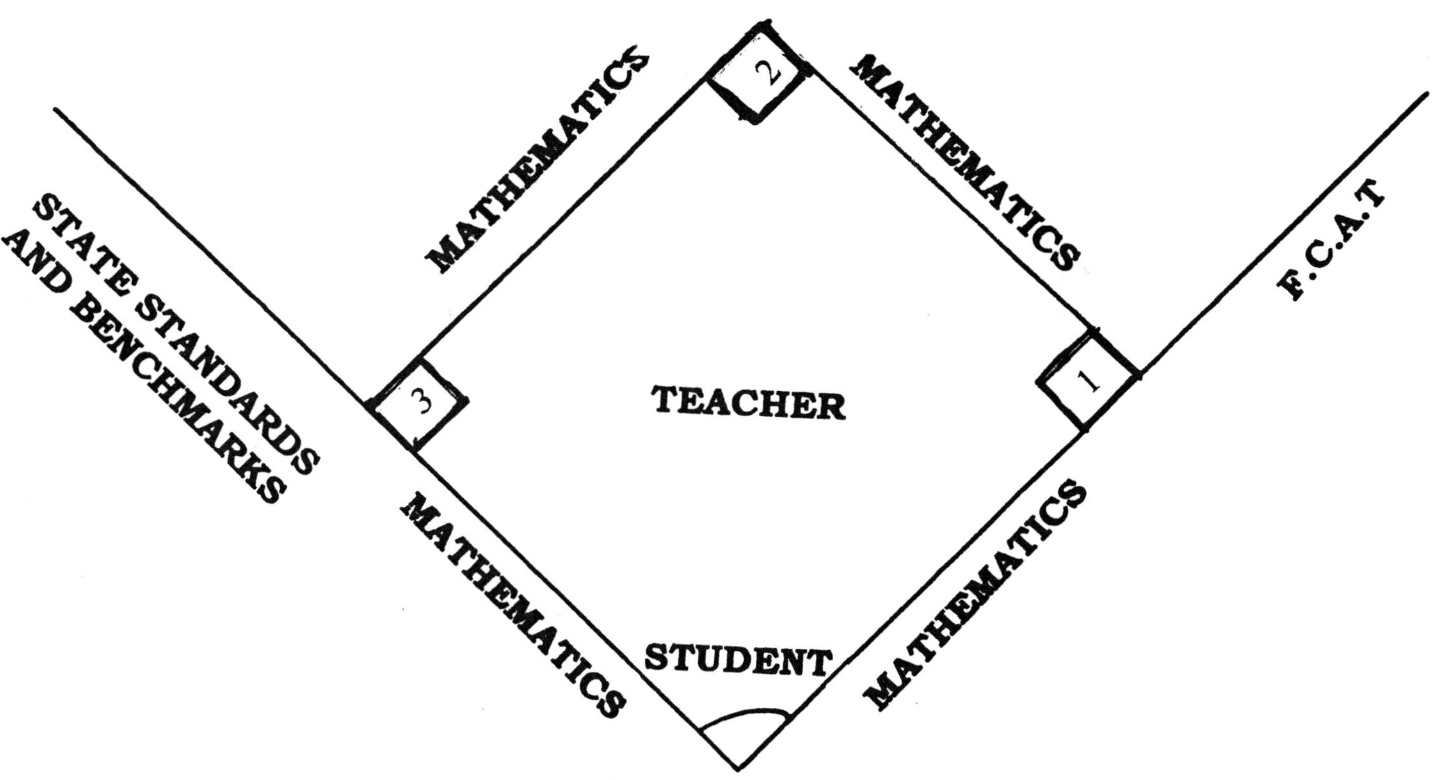

ASSUMPTIONS

This curriculum is based on the following assumptions:

- Children learn best when activities are presented that are appropriate for their age and developmental readiness.
  The lesson plans are presented for each grade from K-2.
- Children learn best when academic concepts are related to a motivating social theme.
  The game of baseball has been the United States national pastime and can serve as a personal connection to learning.
- Children learn best when motor learning and movement are involved in the process.
  Many activities presented are hands-on/movement-involved.
- The state curriculum standards and benchmarks are the foundation of each county's overall curriculum plan.
  This guide follows the state curriculum standards and benchmarks for mathematics. .
- Teachers are provided with the flexibility of developing their own methodology to implement the state and county curriculum standards.
  The game of baseball serves as a foundation for instruction.
- Students learn best when instruction is presented through the emotions of enjoyment, fun and personal experience.
  Discussions and learning activities are predicated on the students' experiences and enjoyment of the game of baseball.

ASSESSMENT

DATA SURVEY INSTRUMENT

Pre: To determine the students awareness, interest, participation and experiences with the game of baseball or softball. The following questions could be asked in a class discussion to determine the level of achievement of the above objective. The questions could be asked and the number of students responding recorded for future discussion, and as an introduction to teaching the lessons.

Participation: These questions will determine the extent of students' direct involvement in and motivation to learn academics through the game of baseball or softball. A class list of student names and the number of the following questions across the top of the page can serve as a record of responses.

1. How many students have ever played the game of baseball or softball on a Little League team? Boys______ Girls ______
2. Who is currently playing on a team now? Boys ______ Girls ______
3. Who has a brother, sister or parent or relative who has played baseball or softball in high school or college, or semi-professionally or professionally?
4. Who has been to a professional baseball game? Name the team!
5. Can anyone name professional baseball teams and their cities?
6. Can anyone name the various positions that players play on the field?
7. Can anyone explain what an inning is?
8. Can anyone name the bases a batter runs to after hitting the ball?
9. What does an umpire do?
10. Who knows what a single, double, triple, and home run are?

These questions will spark the students' motivation for learning through this curriculum as well as provide the teacher with background information regarding the students' knowledge of baseball.

ORGANIZATION FOR EFFECTIVE INSTRUCTION

Lesson Highlights:

The format of each lesson includes the state curriculum standards and benchmarks and sub-topic for the specific activity, the equipment needed for the lesson, the pre-activity discussion, the activity, follow-up discussion, assessment, variations, worksheet and appendix. Teachers are encouraged to review all aspects of the lesson format prior to beginning. Previous lessons and lessons to come are always motivating for students along with tying the standards together in strands. The worksheets and appendix are also indicated as they relate to the specific lesson.

Remember, your enthusiasm for the inclusion of baseball as the vehicle for learning will be a home run for the students.

METHODOLOGY

This curriculum guide is presented with a focus on safeguarding each teacher's inherent professional right to be creative along with the individual development of his/her teaching styles. The activities and discussions presented reserve the absolute choice of the teacher to determine their best practices in the teaching-learning process. This guide is being presented toward that end.

The length of time devoted to each lesson can be varied depending on the student's interest, involvement, experiences, time of day and grade level. Data for the development of this curriculum was collected through:

- Interviews with teachers and administrators
- An analysis of various other curriculum guides and procedures
- An evaluation of what motivates students to learn
- Introspection of many years of teaching experience
- An analysis of demographic areas relating to the sport of baseball
- An analysis of best instructional practices
- An analysis of curriculum and staff development opportunities
- Classroom observations

MOTIVATING LEARNERS

Students will be more motivated to learn academic concepts if they are presented through the game of baseball. Their direct experiences in the game of baseball or softball can be an invaluable source of motivation toward learning. An enthusiastic teacher breeds enthusiastic learners. Using aspects of the game of baseball as the foundation to the academic areas of the state curriculum standards and benchmarks could also encourage some students who have never played the game to possibly try out for a Little League team. Learning can and should be fun, exciting and motivating. The great American pastime of baseball can offer this opportunity to the teacher as well as students.

OPTIONS FOR INDIVIDUALIZING

Among all of the learning styles each child possesses, the most basic is through the tactile-kinesthetic. Most children in grades kindergarten through second do not have the ability to cognitively manipulate thought processes when learning academic concepts. They therefore rely on concrete manipulatives. Children also relate memories of physical experiences to their learning process. If they have played or have fantasized about playing baseball or softball, this will serve as a motivation to enhance whatever learning style becomes dominant.

The pre-discussion portion of each lesson is critical to serve as the trigger to individualize their past experiences and assist in learning the benchmark concept. Pre-determining students' experiences with the game of baseball or softball will serve the teacher with a written record of each child's reference point, which can be drawn upon during the activity portion of the lesson.

PLANNING TIPS

Plan to use the lesson plans from this supplemental curriculum throughout the school year. As each Big Idea and specific benchmark are taught, support the concepts with the baseball activities in these lessons. Utilize the variations suggested as additional reinforcement of the benchmark concepts.

Review the whole curriculum, collect equipment for the specific lesson, look over the accompanying worksheet and review the specific appendix related to the lesson.

Motivate the class for each lesson by relating the activity to children's actual experience when appropriate. Review the data collected in the Assessment/Data Survey Instrument.

Review all of the state curriculum standards and benchmarks before beginning on an individual one. Review all of the benchmarks before teaching a lesson from a specific benchmark.

Relate and/or review previous lessons to the one that you are about to teach.

Relate one subject area to another through integration.

Coordinate special subject areas to each lesson where applicable.

LEARNING/COGNITIVE DISABLED STUDENTS

Students with mild to severe learning disabilities are addressed on the pages containing the Math Standards, Big Ideas, and benchmarks. Below the benchmarks for each grade are Access Points for Students with Significant Cognitive Disabilities.

Three levels of disability are categorized as Independent, Supported and Participatory. The activities and variations listed on the lesson plan pages should be modified accordingly to accommodate these students.

Students with learning disabilities may become especially motivated to learn mathematic benchmarks because of the unique nature of incorporating the game of baseball.

PROVIDING INSTRUCTIONAL SUPPORT FOR TEACHERS

I. This curriculum:

- ❖ Reflects a coherent design including main ideas and essential questions clearly guiding the design of, and alignment with, assessments, teaching and learning activities.
- ❖ Makes clear distinctions between main ideas and essential questions, and the knowledge and skills necessary for learning.
- ❖ Incorporates instruction and assessment that reflects the facets of understanding—the design provides opportunities for students to explain, interpret, apply, shift perspective, empathize, and self-assess.
- ❖ Anchors assessment of understanding with authentic performance tasks calling for students to demonstrate their understanding and apply knowledge and skills.
- ❖ Uses clear criteria and performance standards for teacher, peer, and self-evaluations of student knowledge.
- ❖ Enables students to revisit and rethink important ideas to enhance their understanding.

II. The curriculum will help the teacher:

- ❖ To inform students of the main ideas, essential questions, performance requirements, and evaluative criteria at the beginning of each Sunshine State Standard.
- ❖ Attract students' interest through the game of baseball while students examine and explore learning.
- ❖ Use a variety of strategies to promote deeper understanding of subject matter.
- ❖ Facilitate students' involvement rather than simply telling.

- ❖ Use questioning, probing, and feedback to stimulate student reflection and rethinking.
- ❖ Use information from ongoing assessments as feedback to guide instruction.

If the entire elementary school adopts this curriculum, the students will realize a continuity from one subject area to another and one grade level to another. The lessons will be more meaningful as their awareness and knowledge of baseball develops.

E.S.O.L./E.S.E. ACCOMODATIONS

The following general accommodations or modifications are presented for the teacher to consider in the direct instruction phase of each lesson. Specific individual needs should be addressed utilizing the suggested list of items. Others from the child's IEP should also be noted.

Learning Environment:

1. Provide visual aids or graphic organizers
2. Provide study guides/outlines/notes when requiring students to copy from board/overheads
3. Break longer presentations into shorter segments
4. Write key points on board
5. Provide alternate but comparable segments
6. Allow students to use devices or manipulatives
7. Highlight distinctive features or concepts
8. Preferential seating
9. Reduce/minimize visual distractions
10. Reduce/minimize auditory distractions

Social/Emotional Behavior

1. Specialized training in self-advocacy and understanding of exceptionality

2. Implement a behavior management system/level system with weekly assessments
3. Use daily individualized behavior communication with parent
4. Use weekly behavior communication with parent with agreed upon reinforcements and consequences
5. Opportunity to take time-outs

Study Skills/Independent Learning:

1. Instruction in study skills
2. Organizational skills (i.e., notebook, workspace, desk, etc.)
3. Provide periodic check of work
4. Utilize homework assignment notebook/agenda
5. Provide written timelines for long-term assignments
6. Instruction in time management

Communications:

1. Cueing or reminders regarding appropriate communication (body language)
2. Reduce verbal information
3. Check for comprehension beyond what is normally provided for students
4. Elicit responses; help students "think" through the questions/task
5. Reframe inappropriate dialogue in a more socially appropriate manner

BASEBALL EXERCISES FOR THE CLASSROOM

These exercises can be utilized in a variety of ways in the classroom setting. The most important aspect is to use them on a regular basis whether in the A.M. as part of the opening exercises, or as an introduction to the specific lesson. Selecting children as leaders can be another motivating use of the activity. Incorporating the exercises on a daily basis can also develop the physical strength and flexibility of the children.

Exercises can also be used at any time during the instructional day. The teacher should determine the need for a change of pace during learning. An especially appropriate time is following long periods of concentration or sitting. A change of pace is vital with younger children during their school day. The exercises that follow were designed to be accomplished in a limited space in the classroom. As for baseball players, the exercises focus on strength and flexibility.

CLASSROOM EXERCISES

DEVELOPING FLEXIBILITY

Shoulder:
1. Starting position: Straddle, arms raised sideways to shoulder level, pull arms slightly backward. Repeat this movement several times in a comfortable rhythm and slowly.

Variations: Arms raised a little bit above the shoulder level. As you move the arms back and forth, keep the straight body posture. Avoid excessive motion of any other body parts.

- Raise arms 45 degrees
- Raise arms straight, overhead
- Repeat with arms bent

2. Starting position: Straddle legs, one arm hanging, the other raised upward. Reverse the positions of the arms. The lower arm may move backward a little. Start slowly. Do not move arms behind body.

3. Starting position: Straddle, arms raised sideways. Pull both arms backward a little bit—move both arms in small circles backward—emphasize the backward motion. This movement also strengthens the shoulders.

Variations:

- Repeat with arms raised and at sides.
- Change directions.
- Move one arm forward and the other backward.
- Increase circle size.

Spine and Pelvic Area:

The spine consists of 33 vertebrae arranged in a special S shape which helps protect the nerves inside of it and the pathways to the brain. Between every two vertebrae there is a joint that permits the movement. Since each joint permits a little motion, the total result is the bending of the spine; and indeed, we can bend the spine in all directions and combine any bending with a twist. This arrangement is very similar to a spiral spring that can be compressed, bent, and twisted, and can return to original position. The spine can do it, too. When you land after a jump, the elasticity of the spine permits it to be compressed a bit before returning to normal. This makes the jump safe and relatively comfortable. The spine can bend when you pick up a shoe or twist when you turn to answer a call from behind you; this is accomplished by the muscles which cover the spine and the trunk on all four sides.

BENDING FORWARD

1. Starting position: Straddle legs, bend forward and down. Keep legs straight, bob several times slowly. Don't bend to your maximum the first time. Keep legs straight. Do *not* try to reach the floor if it is difficult in the beginning. It will take time before you will reach the floor without undue discomfort.

Variations:

- Stand with a narrow straddle.
- Palms on the floor.

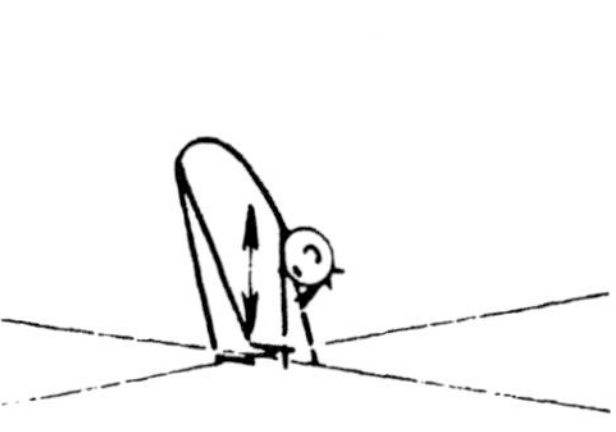

- From a straddle starting position, reach backward with hands between the legs.

- In any of the previous starting positions, bend toward one foot and then toward the other.

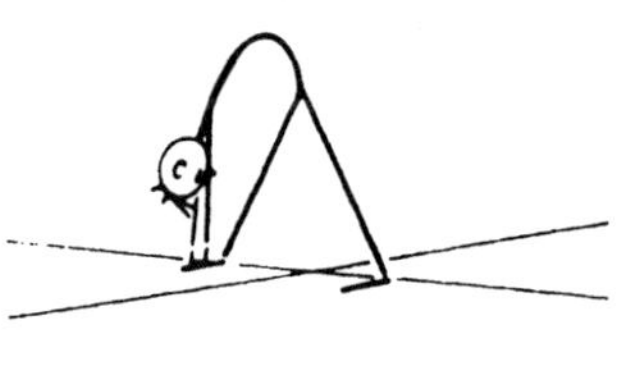

- In straddle position, bend forward and down and grasp your ankles. Pull yourself toward the legs so that your face will approach the knees.

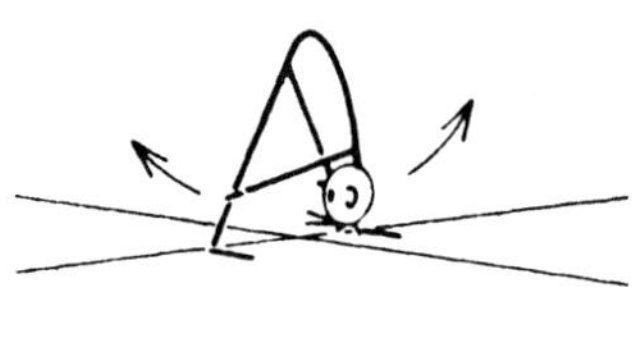

BENDING SIDEWAYS

2. Starting position: Straddle, arms along the sides of the body—bend your trunk to one side and straighten up. Repeat to the other side. Bend *exactly* to the side, not forward or backward.

Variations:

- Bend to the same side and up straight several times.

- Slide hands along legs as you bend to one side. Stay exactly sideways.

- With palms placed behind the neck, pull your elbows slightly backward during the side bending.

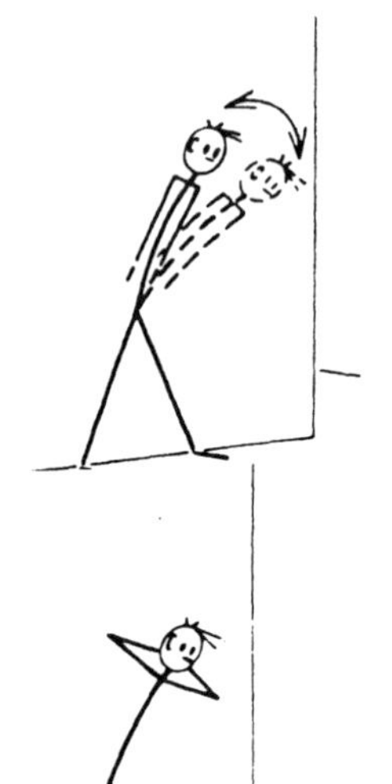

- With hands placed on the head, pull your elbows slightly backward during the side bending. This elevation of the arms raises the center of gravity of the body and increases the degree of difficulty in these movements.

- Repeat with arms raised upward (parallel or in various angles). This movement is quite difficult to perform well and is recommended only to students who have mastered the previous steps in side bending.

- In a straddle position with palms behind neck—twist the trunk to one side and return to the starting position. Let the elbows lead the twist, avoid excessive motion at the hips.

- In order to increase the degree of difficulty of the side bending and the twists, it is necessary to stabilize the pelvic girdle. One way is to perform the movement in a side lunge position.

SITTING
In an L position, legs apart, bend trunk slowly forward and toward one foot. Sit back up straight. Bend toward other foot and up.

Variations:

- Cross legs as you are bending forward.
- Toes flexed upward as bending forward.

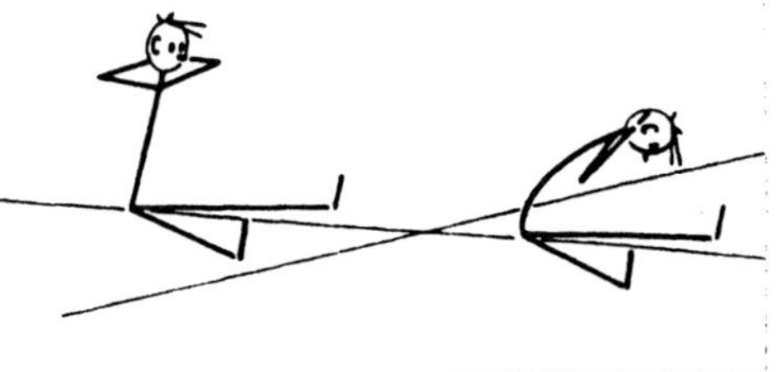

- Place both palms on the floor outside one ankle when you bend toward the other.

DEVELOPING STRENGTH:

Development of strength can be accomplished either by maintaining the position (isometric work or static work), or by moving from one position to another (isotonic work or dynamic work). The determination of the degree of difficulty in the development of strength (isometrically or isotonically) is based on the following for considerations.

Weight can be increased by added external weight or by changes in the initial position.

Duration can be controlled by changing the number of repetitions in the set (a set is a group of repetitions) and by changing the number of sets with a short interval between sets.

The distance or height will vary and help determine difficulty.

The final consideration is the *speed* of the performance.

The following movements are based on these considerations and are classified by the region of the body: shoulder girdle and arms, upper back, abdomen, legs.

TRUNK STRENGTH:

- Lie on back, knees bent slightly, palms on floor at side. Roll upper body up to a sitting position and lie back down.

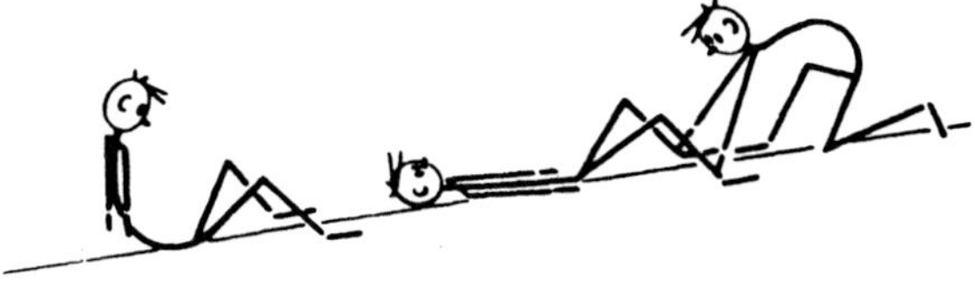

LEG STRENGTH:

- Lie on back, knees slightly bent, straighten one leg up in air, and back to rest on floor. Repeat several times with each leg.

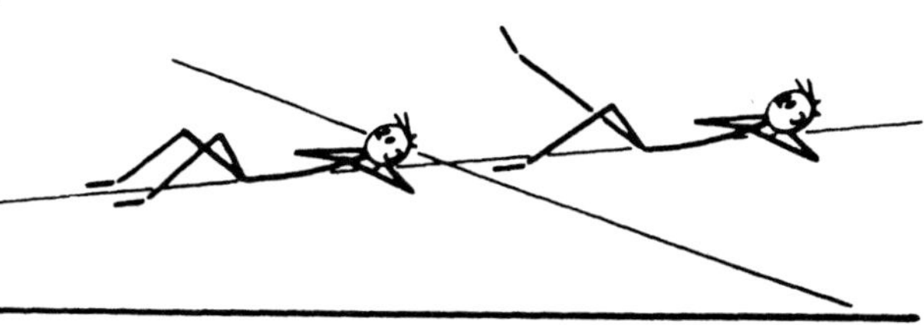

- Lie on back, knees bent up, feet on floor. Straighten one leg and hold off floor six inches. Repeat with each leg. Return to a bent leg position with foot on floor.

- Lean back on elbows in a half-sitting position, knees bent and feet on floor. Straighten both legs as you push on floor with hands.

- Repeat with legs straight and sitting upright. Raise legs off floor one at a time and then both together.

THE SHOULDER AND ARMS

Bear Walk--

Starting position: "On four" facing the floor

1. Walk forward a few steps and back to original starting position.

Variations:

- **Straight arms and straight legs.**

- **Left arm and left leg move simultaneously.**
- **Move in a square—forward, to the left, backward, to the right.**

- Raise the center of the body to a high bridge.
- Push-up position (straight back).

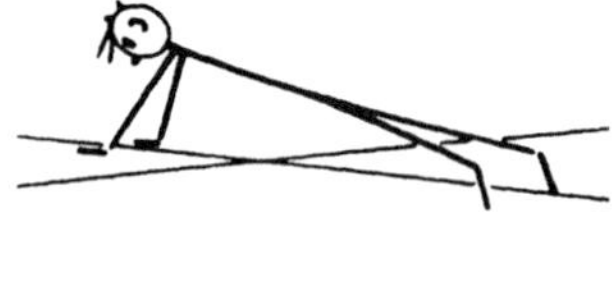

- Change the base of support with the hands. The difficulty will be increased as the hands are moved together, farther apart or forward.

- In push-up position, extend one leg up, back and straighten.

- Arm circles; forward and backward. Larger circles, smaller circles. Slower circles, larger circles.

BACK STRENGTH:

- Straddle stand, hands behind neck. Tilt body forward, keeping back straight. Move back to upright position.

- Tilt forward and lower to a right angle with trunk and legs.

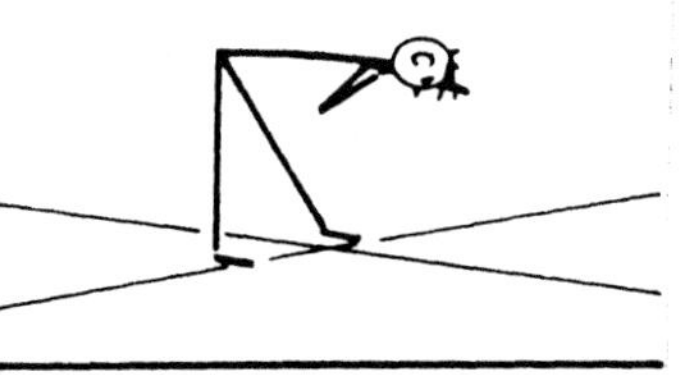

- Extend arms forward while in forward tilted position.

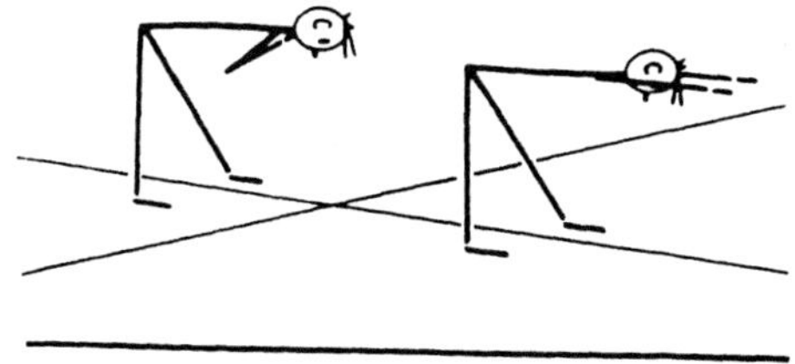

TRUNK STRENGTH:

- Lie on back, knees bent slightly, palms on floor at side. Roll upper body up to a sitting position and lie back down.

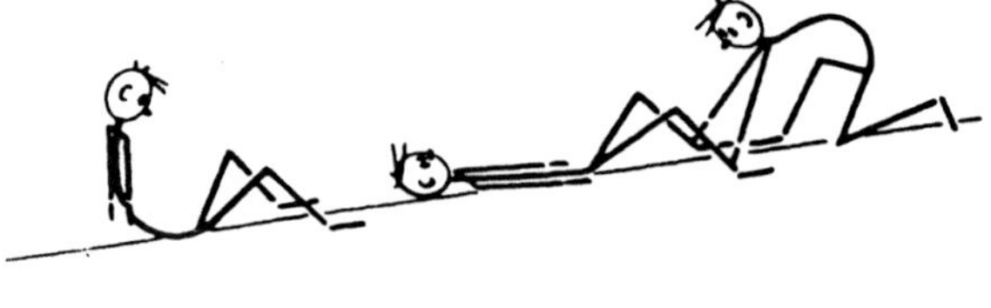

LEG STRENGTH:

- Lie on back, knees slightly bent, straighten one leg up in air, and back to rest on floor. Repeat several times with each leg.

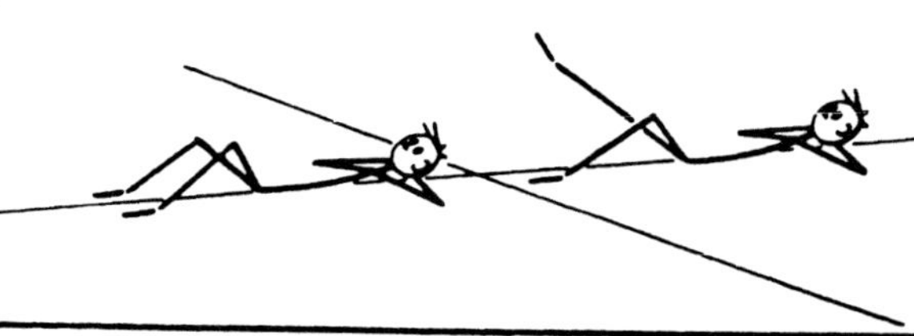

- Lie on back, knees bent up, feet on floor. Straighten one leg and hold off floor six inches. Repeat with each leg. Return to a bent leg position with foot on floor.

- Lean back on elbows in a half-sitting position, knees bent and feet on floor. Straighten both legs as you push on floor with hands.

- Repeat with legs straight and sitting upright. Raise legs off floor one at a time and then both together.

- Start in a standing position. Bend the knees to a half squat, hold and return to a standing position.

- From a standing position, take a long step forward, lunge, bending the knee and return to standing position. Repeat alternating legs.
  Lunge forward, sideward and backward.

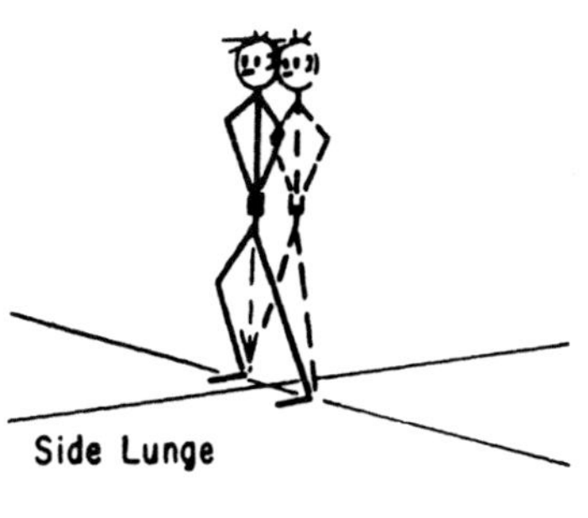

REFERENCE; EXERCISE FIGURES

DEVELOPMENTAL MOVEMENT, MUSKA, MOSSTON, CHARLES E. MERRILL PUBLISHING C0

KINDERGARTEN

FLORIDA MATHEMATICS STANDARDS

K-8 GRADE-LEVEL STANDARDS

Big Ideas

Big Ideas are standards that are aligned with the *Curriculum Focal Points* released by the National Council of Teachers of Mathematics (NCTM). They include standards which should be the primary focus of mathematics instruction for each grade level, K-8. Establishing proficiency with these standards at each successive grade level will prepare a strong foundation for learning mathematics in subsequent grades.

There are three Big Ideas for each grade. The Big Ideas do not address the same topics for each grade, recognizing that at each level there are certain skills which must be honed to prepare students for more rigorous instruction as they move to the next grade. The order of the Big Idea standards does not determine the order of instruction nor does it indicate that one idea requires greater instructional emphasis. The Big Ideas are assigned numbers 1, 2, or 3 without regard to the content in each of them.

Supporting Ideas

Supporting ideas are standards which are fundamental to sound mathematics instruction. Also aligned with the *Curriculum Focal Points*, Supporting Ideas are not less important than the Big Ideas but are key components to a structurally sound mathematics education.

Supporting Ideas are standards that serve one or more of the following purposes:

1) Establishing connections to and between the strands of mathematics as defined by NCTM (Probability has been extracted from Data Analysis and stands alone.);
2) Preparing students for future mathematics teaching and learning by focusing on conceptual understanding of concepts; and
3) Addressing gaps in instruction that may appear insignificant but are important to the understanding, fluency, and application of mathematics ideas to problem solving. The

Kindergarten

Benchmark Coding Scheme

MA.	K.	A.	1.	1
Subject	Grade Level	Body of Knowledge	Big Idea / Supporting Idea	Benchmark

Body of Knowledge Key:

A ~ Algebra

C ~ Calculus

D ~ Discrete Mathematics

F ~ Financial Literacy

G ~ Geometry

P ~ Probability

S ~ Statistics

T ~ Trigonometry

Access Points Coding Scheme

MA.	K.	A.	1.	In.a
Subject	Grade Level	Body of Knowledge	Big Idea / Supporting Idea	Access Point

Access Points Key:

In ~ Independent

Su ~ Supported

Pa ~ Participatory

K-8 MATHEMATICS STANDARDS
GRADE K

BIG IDEA 1: *Represent, compare, and order whole numbers and join and separate sets.*	
BENCHMARK CODE	**BENCHMARK**
MA.K.D.1.1	Represent quantities with numbers up to 20, verbally, in writing, and with manipulatives.
MA.K.D.1.2	Solve problems including those involving sets by counting, by using cardinal and ordinal numbers, by comparing, by ordering, and by creating sets up to 20.
MA.K.D.1.3	Solve word problems involving simple joining and separating situations.

Access Points for Students with Significant Cognitive Disabilities

Independent:	***Supported:***	***Participatory:***
MA.K.D.1.In.a Represent quantities to 5 using sets of objects and number names. MA.K.D.1.In.b Use one-to-one correspondence to count and compare sets of objects to 5. MA.K.D.1.In.c Solve problems with up to 5 objects, involving simple joining (putting together) and separating (taking away) situations.	MA.K.D.1.Su.a Represent quantities to 3 using sets of objects and number names. MA.K.D.1.Su.b Use one-to-one correspondence to count sets of objects to 3. MA.K.D.1.Su.c Solve problems with up to 3 objects involving simple joining (putting together) situations.	MA.K.D.1.Pa.a Respond to a prompt to indicate desire for more of a preferred, familiar action or object. MA.K.D.1.Pa.b Respond to a prompt to indicate desire to stop an action or activity. MA.K.D.1.Pa.c Respond to a counting cue to begin a familiar routine. MA.K.D.1.Pa.d Demonstrate a favorable or positive response to a preferred stimulus.

BIG IDEA 2: *Describe shapes and space.*	
BENCHMARK CODE	**BENCHMARK**
MA.K.G.2.1	Describe, sort and re-sort objects using a variety of attributes such as shape, size, and position.
MA.K.G.2.2	Identify, name, describe and sort basic two-dimensional shapes such as squares, triangles, circles, rectangles, hexagons, and trapezoids.
MA.K.G.2.3	Identify, name, describe, and sort three-dimensional shapes such as spheres, cubes and cylinders.
MA.K.G.2.4	Interpret the physical world with geometric shapes and describe it with corresponding vocabulary.
MA.K.G.2.5	Use basic shapes, spatial reasoning, and manipulatives to model objects in the environment and to construct more complex shapes.

Access Points for Students with Significant Cognitive Disabilities		
Independent:	***Supported:***	***Participatory:***
MA.K.G.2.In.a Sort objects by single attributes, including shape and size. MA.K.G.2.In.b Match and name two-dimensional shapes, including circle and square. MA.K.G.2.In.c Match examples of three-dimensional objects, such as balls (spheres) and blocks (cubes). MA.K.G.2.In.d Identify shapes, including circle and square, in the environment. MA.K.G.2.In.e Identify spatial relationships, including in, out, up, down, top, bottom, on, and off.	MA.K.G.2.Su.a Identify square objects or pictures when given the name. MA.K.G.2.Su.b Identify three-dimensional objects, such as a block (cube) or ball (sphere). MA.K.G.2.Su.c Identify square shapes in the environment when given the name. MA.K.G.2.Su.d Identify spatial relationships, including on, off, up, and down.	MA.K.G.2.Pa.a Respond to a prompt to identify a familiar three-dimensional object in a familiar routine. MA.K.G.2.Pa.b Respond to one directional prompt in a familiar routine.

BIG IDEA 3: *Order objects by measurable attributes.*

BENCHMARK CODE	BENCHMARK
MA.K.G.3.1	Compare and order objects indirectly or directly using measurable attributes such as length, height, and weight.

Access Points for Students with Significant Cognitive Disabilities

Independent:	*Supported:*	*Participatory:*
MA.K.G.3.In.a Compare overall size and length of objects and describe using terms such as big, small, long, and short.	MA.K.G.3.Su.a Identify size of objects using terms such as big and little.	MA.K.G.3.Pa.a Respond to differences in familiar persons, actions, or objects within a familiar routine.

SUPPORTING IDEAS

Algebra

BENCHMARK CODE	BENCHMARK
MA.K.A.4.1	Identify and duplicate simple number and non-numeric repeating and growing patterns.

Access Points for Students with Significant Cognitive Disabilities

Independent:	*Supported:*	*Participatory:*
MA.K.A.4.In.a Match two-element repeating patterns of sounds, physical movements, and objects.	MA.K.A.4.Su.a Match identical sounds, physical movements, and objects.	MA.K.A.4.Pa.a Demonstrate distinctive responses to preferred vs. non-preferred stimuli in a familiar routine.

SUPPORTING IDEAS

Geometry and Measurement

BENCHMARK CODE	BENCHMARK
MA.K.G.5.1	Demonstrate an understanding of the concept of time using identifiers such as morning, afternoon, day, week, month, year, before/after, and shorter/longer.

Access Points for Students with Significant Cognitive Disabilities

Independent:	*Supported:*	*Participatory:*
MA.K.G.5.In.a Identify concepts of time, including day, night, morning, and afternoon, by relating activities to a time period.	MA.K.G.5.Su.a Identify concepts of time, including day and night, by relating daily events to a time period.	MA.K.G.5.Pa.a Respond to the environmental cue for a preferred activity within a regularly scheduled routine.

LESSON PLAN

SUBJECT: MATH
GRADE: K

BIG IDEA #1: Represent, compare, and order whole numbers, and join and separate sets.

BENCHMARK #1: (MA.K.D.1.1) Represents quantities with numbers.

TITLE OF LESSON: Counting Different Ways

EQUIPMENT/MATERIALS: crayons, scissors, concrete objects: bat, ball, other baseball equipment.

PRE-DISCUSSION: Team equipment manager obtains items needed for a game. Explain what a manager's role is.

ACTIVITY: Teacher discusses that whole numbers can be represented in different ways. Students color and cut out worksheet puzzle and reassemble by pasting on another sheet of paper.

VARIATIONS: (1) Small group activity; cut out, mix up, and work cooperatively to reassemble. Use up to the number ten (10).

(2) Use different colored paper for each number.

ASSESSMENT: Successful completion of puzzles.

WORKSHEET: Bat, one, 1, first (4-piece puzzle on 8x11 paper)

APPENDIX: (C) Equipment

WORKSHEET

Subject: Math
Grade: K
Big Idea #1
Benchmark #1

____
Student Name

Counting Different Ways

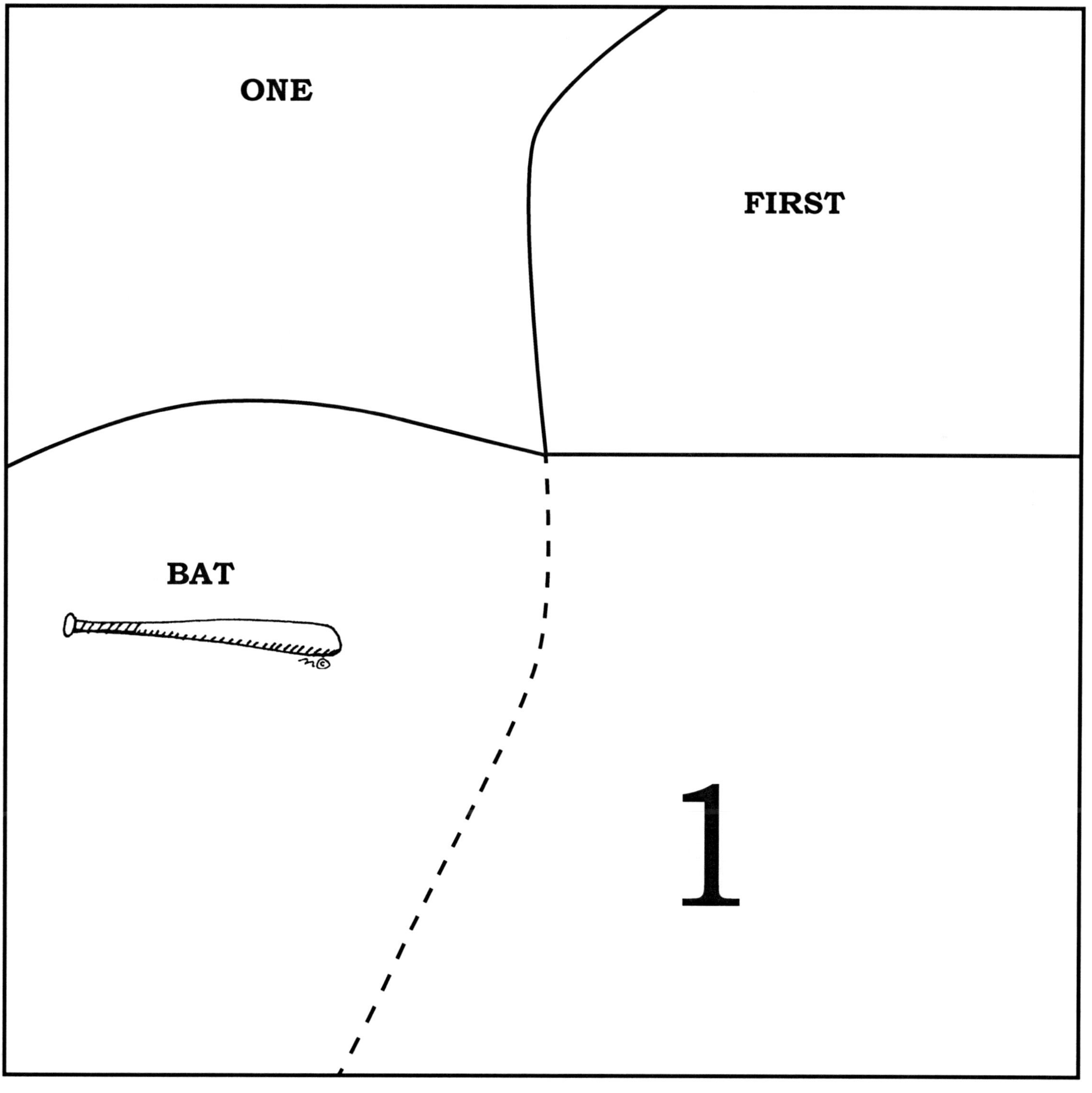

LESSON PLAN

SUBJECT: MATH
GRADE: K

BIG IDEA #1: Represent, compare, and order whole numbers, and join and separate sets.

BENCHMARK #2: (MA.K.D.1.1) Solves problems by counting, using cardinal and ordinal numbers, comparing, ordering, and creating sets.

TITLE OF LESSON: Count the Runs And Be A Scorekeeper

EQUIPMENT/MATERIALS: Manipulatives

PRE-DISCUSSION: What was the score of a game you played in? Relate to the students real life situtations.

ACTIVITY: Announce the score of a sample game: Tampa Bay—16, New York Yankees—15. Students count the number of runs Tampa Bay scored by 2s and 4s and the number of runs the New York Yankees scored by 5s.

VARIATIONS: Which numbers could not be used to count to the Tampa Bay score without a remainder? (3,5,6,7,9,10, etc.) Change the team names and runs scored to count by different numbers.

ASSESSMENT: Record the names of students able to count correctly.

LESSON PLAN

SUBJECT: MATH
GRADE: K

BIG IDEA #1: Represent, compare, and order whole numbers and join and separate sets.

BENCHMARK #3: (MA.K.D.1.3) Solves word problems involving simple joining and separating situations.

TITLE OF LESSON: Select Your Team

EQUIPMENT/MATERIALS: Inch cubes or magnetic figures, scissors, magazines, newspapers.

PRE-DISCUSSION: Review player positions on field. For teams to be fair, the number of players should be equal.

ACTIVITY: Teacher draws (2) two playing fields on large paper or chalkboard. Teacher adds or removes manipulatives from playing fields to represent none, more than, same, one more than, etc.

VARIATIONS: Cut out pictures of baseball players from the newspaper or magazines. Add and subtract pictures for counting.

ASSESSMENT: Correct student responses and teacher observations.

WORKSHEET: Use baseball cards if available.

APPENDIX: (B) Baseball Field and Positions

LESSON PLAN

SUBJECT: MATH
GRADE: K

BIG IDEA#2: Describes shapes and space.

BENCHMARK #1: (MA.K.G.2.1) Describes, sorts, and re-sorts objects using a variety of attributes.

TITLE OF LESSON: Sort Your Baseball Equipment

EQUIPMENT/MATERIALS: Hat, bat, ball, glove, shoe base, uniform, etc.

PRE-DISCUSSION: Identify descriptive words that can describe various baseball equipment.

ACTIVITY: Children describe baseball equipment using words in pre-discussion. Children can identify and sort equipment by using the worksheet. Sort equipment by size, shape or use.

VARIATIONS: Students draw and sort their own pictures of equipment.

ASSESSMENT: Students complete worksheet correctly.

WORKSHEET: Follow instructions on worksheet after students color and cut out items.

APPENDIX: (C) Equipment

WORKSHEET

Subject: Math
Grade: K
Big Idea #2
Benchmark #1

____
Student Name

Sort Your Equipment (#1 of 2)

Students color and cut out each equipment item. Students sort items on their desk: those used for safety, those to be worn, those used to score a run, etc.

Students place items in spatial relation to each other on their desk:

- Place the ball under the cap,
- Place the glove above the shoe,
- Place the bat beside the glove,
- Also use inside, next to, below, apart, etc.

Students place objects together and describe their spatial relation.

WORKSHEET

Subject: Math
Grade: K
Big Idea #2
Benchmark #1

____
Student Name

<u>Sort Your Equipment</u> (#2 of 2)

BALL

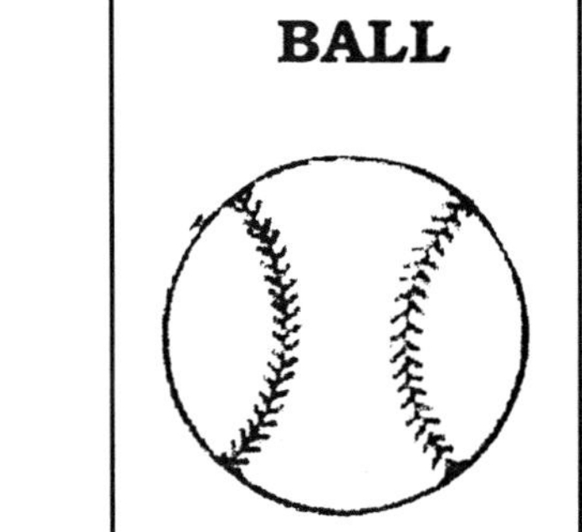

BAT

GLOVE

BASE

HELMET

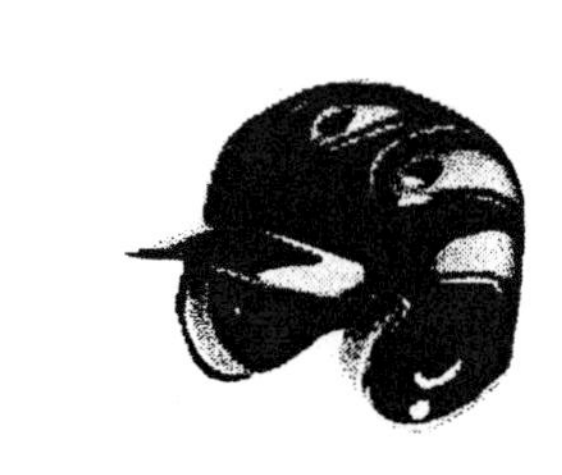

UNIFORM

LESSON PLAN

SUBJECT: MATH
GRADE: K

BIG IDEA #2: Describes shapes and space.

BENCHMARK #2: (MA. K.G.2.2) Identifies, names, describes and sorts two-dimensional and shapes.

BENCHMARK #3: Identifies, names, describes and sorts three-dimensional shapes.

TITLE OF LESSON: Guess and Tell

EQUIPMENT/MATERIALS: Bat, ball, glove, scissors, paper

PRE-DISCUSSION: Show students a picture of and the actual object. Discuss the difference.

ACTIVITY: Have the students feel the objects and then tell what is different about the picture versus the object. Have students make a baseball by crumpling up a paper. Have students roll up and tape newspaper to represent the size of a bat.

VARIATIONS: Students describe and name cubes, spheres and cylinders that are baseball objects. Use a book from the library to show pictures.

ASSESSMENT: Students can identify two- and three-dimensional objects in their environment.

LESSON PLAN

SUBJECT: MATH
GRADE: K

BIG IDEA #2: Describes shapes and space.

BENCHMARK #4: (MA.K.G.2.4) Interprets physical world with geometric shapes and describes it with corresponding vocabulary.

BENCHMARK #5: (MA.K.G.2.5) Uses basic shapes, spatial reasoning and manipulatives to model objects in the environment and to construct more complex shapes.

TITLE OF LESSON: Baseball Shapes

EQUIPMENT/MATERIALS: Ball, base, home plate, diagram of a field.

PRE-DISCUSSION: Students name shapes found in the game of baseball.

ACTIVITY: Students identify shapes: ball—sphere, base—square, pitchers mound—rectangle, home plate—pentagon, and base path—square.

VARIATIONS: Walk class to a baseball field to find actual shapes.

ASSESSMENT: Students can identify other shapes in their environment.

WORKSHEET: Students color and cut out shapes.

APPENDIX: (B) Playing Field

WORKSHEET

Subject: Math
Grade: K
Big Idea #2
Benchmark #4 & 5

____
Student Name

Baseball Shapes

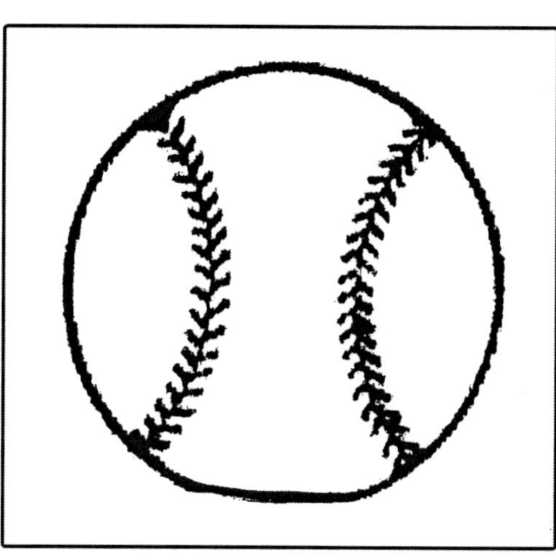

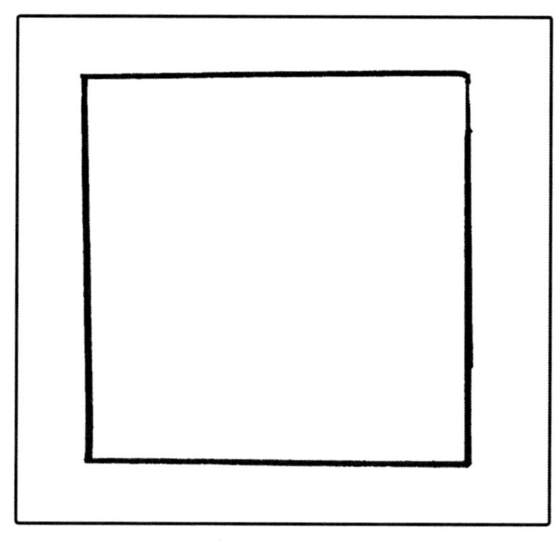

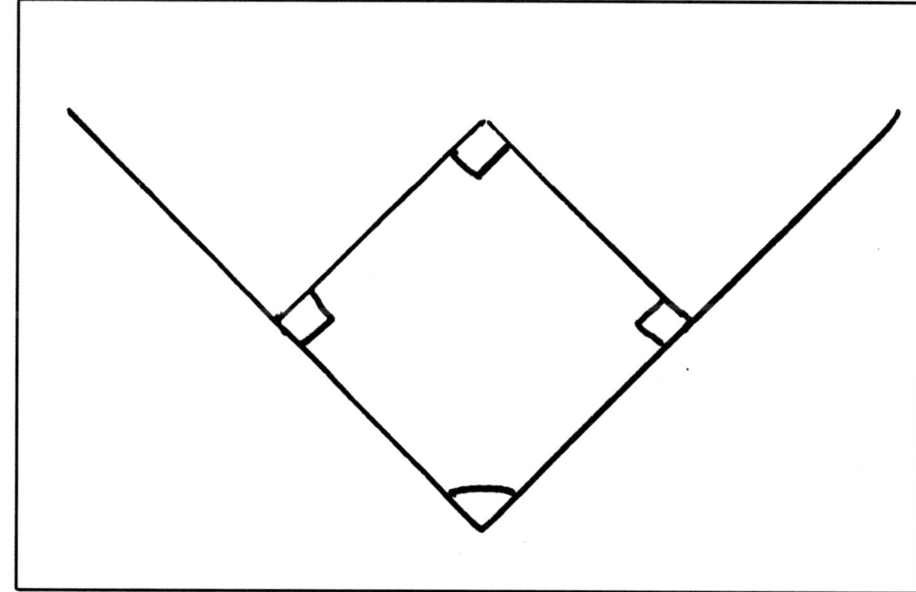

LESSON PLAN

SUBJECT: MATH
GRADE: K

BIG IDEA #3: Orders objects by measurable attributes.

BENCHMARK #1: (MA.K.G.3.1) Compares and orders objects indirectly or directly using measurable attributes.

TITLE OF LESSON: How Much Does Your Baseball Equipment Weigh?

EQUIPMENT/MATERIALS: Baseball, glove, cleated shoe and bats

PRE-DISCUSSION: Discuss various equipment used in a game. Review various glove, bat and shoe sizes. What size would you use?

ACTIVITY: Students estimate the weight of a baseball, glove, cleated shoe and bat. Place them in order from lightest to heaviest and heaviest to lightest.

VARIATIONS: Use different types of balls--golf ball, playground ball, softball, beachball, etc.--to order their weight.

ASSESSMENT: (1) Have students write or draw their estimate of which objects weigh the most and the least. (2) Have a student lie on the floor and measure their height. Which baseball objects could you use touching together to equal the student's measurement? A bat, shoe and ball; or a bat, glove and shoe; etc.

APPENDIX: (C) Baseball Equipment; (N) Bat Sizes

LESSON PLAN

SUBJECT: MATH
GRADE: K

SUPPORTING IDEA: ALGEBRA

BENCHMARK #1: (MA.K.A.4.1) Identify and duplicate simple number and non-numeric repeating and growing patterns.

TITLE OF LESSON: Be A Scorekeeper

EQUIPMENT/MATERIALS: Manipulatives

PRE-DISCUSSION: What was the score of a game that you played in?

ACTIVITY: Announce the score of a sample game: Tampa Bay—16, New York Yankees—15. Students count the number of runs Tampa Bay scored by 2s and the number of runs the New York Yankees scored by 3s and 5s.

VARIATIONS: Which numbers could not be used to count to the Tampa Bay score without a remainder? (3, 5, 6, 7, 9, 10, etc.)

ASSESSMENT: Record the names of students able to count correctly.

LESSON PLAN

SUBJECT: MATH
GRADE: K

SUPPORTING IDEA: GEOMETRY AND MEASUREMENT

BENCHMARK #1: (MA.K.A.4.1) Demonstrates an understanding of the concept of time.

TITLE OF LESSON: A Day in the Life of a Baseball Player

EQUIPMENT/MATERIALS: Crayons and paper. Divide paper into three sections.

PRE-DISCUSSION: Describe your day by order of events; morning, afternoon, and evening.

ACTIVITY: Guess what a baseball player's day is like. What would the player do in the morning, afternoon, and night? Would they exercise, play catch, practice batting, or play a practice game? What would players in different positions do differently?

VARIATION: How long is a game compared to the student's day? How would the daily schedule differ for a day or night game? List different activities in the daily life of a baseball player during the season. Have the students label morning, afternoon or evening for each activity.

WORKSHEET: Draw a picture of a baseball player exercising, practicing skills, and playing on a field, during the day and a night.

FIRST GRADE

K-8 MATHEMATICS STANDARDS
GRADE 1

BIG IDEA 1: *Develop understandings of addition and subtraction strategies for basic addition facts and related subtraction facts.*

BENCHMARK CODE	BENCHMARK
MA.1.A.1.1	Model addition and subtraction situations using the concepts of "part-whole," "adding to," "taking away from," "comparing," and "missing addend."
MA.1.A.1.2	Identify, describe, and apply addition and subtraction as inverse operations.
MA.1.A.1.3	Create and use increasingly sophisticated strategies, and use properties such as Commutative, Associative and Additive Identity, to add whole numbers.
MA.1.A.1.4	Use counting strategies, number patterns, and models as a means for solving basic addition and subtraction fact problems.

Access Points for Students with Significant Cognitive Disabilities

Independent:	*Supported:*	*Participatory:*
MA.1.A.1.In.a Identify the meaning of addition as adding to and subtraction as taking away from.	MA.1.A.1.Su.a Demonstrate understanding of the meaning of joining (putting together) and separating (taking apart) sets of objects.	MA.1.A.1.Pa.a Respond to the arrival of a familiar person or addition of a familiar object in a routine.
MA.1.A.1.In.b Use counting and one-to-one correspondence as strategies to solve addition facts with sums to 10 and related subtraction facts represented by numerals with sets of objects and pictures.	MA.1.A.1.Su.b Use one-to-one correspondence as a strategy for solving simple number stories involving joining (putting together) and separating (taking apart) with sets of objects to 5.	MA.1.A.1.Pa.b Respond to the departure of a familiar person or removal of a familiar object in a routine.

BIG IDEA 2: ***Develop an understanding of whole number relationships, including grouping by tens and ones.***	
BENCHMARK CODE	**BENCHMARK**
MA.1.A.2.1	Compare and order whole numbers at least to 100.
MA.1.A.2.2	Represent two digit numbers in terms of tens and ones.
MA.1.A.2.3	Order counting numbers, compare their relative magnitudes, and represent numbers on a number line.

Access Points for Students with Significant Cognitive Disabilities		
Independent:	***Supported:***	***Participatory:***
MA.1.A.2.In.a Compare and order numbers 1 to 10. MA.1.A.2.In.b Use one-to-one correspondence to count sets of objects or pictures to 10. MA.1.A.2.In.c Represent numbers to 10 using sets of objects and pictures, number names, and numerals.	MA.1.A.2.Su.a Use one-to-one correspondence to compare sets of objects to 5. MA.1.A.2.Su.b Use one-to-one correspondence to count sets of objects to 5 arranged in a row. MA.1.A.2.Su.c Represent quantities to 5 using sets of objects and number names.	MA.1.A.2.Pa.a Respond to a prompt to indicate desire for more of two or more preferred actions or objects in a familiar routine. MA.1.A.2.Pa.b Respond to a prompt to indicate desire to stop two or more actions in a familiar routine. MA.1.A.2.Pa.c Respond to a counting cue to begin two or more familiar routines.

BIG IDEA 3: ***Compose and decompose two-dimensional and three-dimensional geometric shapes.***	
BENCHMARK CODE	**BENCHMARK**
MA.1.G.3.1	Use appropriate vocabulary to compare shapes according to attributes and properties such as number and lengths of sides, and number of vertices.
MA.1.G.3.2	Compose and decompose plane and solid figures, including making predictions about them, to build an understanding of part-whole relationships and properties of shapes.

Access Points for Students with Significant Cognitive Disabilities		
Independent:	***Supported:***	***Participatory:***
MA.1.G.3.In.a Sort and describe two-dimensional shapes by single attributes such as number of sides and straight or round sides. MA.1.G.3.In.b Identify examples of three-dimensional objects, including sphere and cube. MA.1.G.3.In.c Combine two shapes to make another shape and identify the whole-part relationship. MA.1.G.3.In.d Describe spatial relationships, including over, under, front, back, and between.	MA.1.G.3.Su.a Match common two-dimensional objects by shape, including square and circle. MA.1.G.3.Su.b Name two-dimensional shapes, including circle and square. MA.1.G.3.Su.c Sort common two- and three-dimensional objects by size, including big and little. MA.1.G.3.Su.d Identify spatial relationships, including in, out, top, and bottom.	MA.1.G.3.Pa.a Respond to a prompt to identify a familiar object with a two-dimensional shape, such as circle or square in familiar routines. MA.1.G.3.Pa.b Respond to a prompt to identify two or more familiar three-dimensional objects in familiar routines. MA.1.G.3.Pa.c Demonstrate awareness of one discrete location (area) in the learning environment. MA.1.G.3.Pa.d Respond to two directional prompts in familiar routines.

SUPPORTING IDEAS

Algebra

BENCHMARK CODE	BENCHMARK
MA.1.A.4.1	Extend repeating and growing patterns, fill in missing terms, and justify reasoning.

Access Points for Students with Significant Cognitive Disabilities

Independent:	*Supported:*	*Participatory:*
MA.1.A.4. In.a Match a two-element repeating visual pattern.	MA.1.A.4.Su.a Match objects by single attributes such as color, shape, or size.	MA.1.A.4.Pa.a Indicate anticipation of next step in a familiar routine or activity.

SUPPORTING IDEAS

Geometry and Measurement

BENCHMARK CODE	BENCHMARK
MA.1.G.5.1	Measure by using iterations of a unit and count the unit measures by grouping units.
MA.1.G.5.2	Compare and order objects according to descriptors of length, weight and capacity.

Access Points for Students with Significant Cognitive Disabilities

Independent:	*Supported:*	*Participatory:*
MA.1.G.5.In.a Measure length of objects using nonstandard units of measure and count the units. MA.1.G.5.In.b Compare objects by concepts of length, using terms like longer, shorter, and same; and capacity, using terms like full and empty. MA.1.G.5.In.c Identify concepts of time, including before, after, and next, by relating daily events to a time period.	MA.1.G.5.Su.a Measure length of objects using nonstandard units of measure. MA.1.G.5.Su.b Compare objects by length using terms like long and short. MA.1.G.5.Su.c Identify the concepts of time, including morning and afternoon, by relating daily events to a time period.	MA.1.G.5.Pa.a Respond to differences in familiar persons, actions, or objects in two or more familiar routines. MA.1.G.5.Pa.b Respond to the environmental cue for preferred activities within regularly scheduled routines.

SUPPORTING IDEAS

Number and Operations

BENCHMARK CODE	BENCHMARK
MA.1.A.6.1	Use mathematical reasoning and beginning understanding of tens and ones, including the use of invented strategies, to solve two-digit addition and subtraction problems
MA.1.A.6.2	Solve routine and non-routine problems by acting them out, using manipulatives, and drawing diagrams

Access Points for Students with Significant Cognitive Disabilities

Independent:	*Supported:*	*Participatory:*
MA.1.A.6.In.a Solve real-world problems involving addition facts with sums to 10 and related subtraction facts using numerals with sets of objects and pictures.	MA.1.A.6.Su.a Solve real-world problems involving simple joining (putting together) and separating (taking apart) situations with sets of objects to 5.	MA.1.A.6.Pa.a Demonstrate distinctive responses to preferred vs. non-preferred stimuli in two familiar routines.

LESSON PLAN

SUBJECT: MATH
GRADE: 1

BIG IDEA #1: Develop an understanding of addition and subtraction strategies for basic addition facts and related subtraction facts.

BENCHMARK #1: (MA.1.A.1.1) Model addition and subtraction situations using the concepts of "part-whole," "adding to," "taking away from," "comparing," and "missing addend."

TITLE OF LESSON: What Is Your Team's Score?

EQUIPMENT/MATERIALS: Inch cubes, blocks, manipulatives.

PRE-DISCUSSION: Nine innings are in a baseball game with each team having a chance to score runs. Draw an inning chart.

ACTIVITY:	1	2	3	4	5	6	7	8	9	Total
Team	0	1	0	3	0	2	1	3	__	11
Team	1	2	0	1	0	1	0	2	0	7

Students complete the chart together giving total runs after each inning.

VARIATIONS: (1) What would the total runs scored be if the team had scored 3 runs in the ninth inning? Choose various innings. (2) Which team score is "less than" or "greater than" after each inning? (3) Complete the missing runs scored to get the total runs.

ASSESSMENT: Students complete the worksheet correctly.

WORKSHEET: What Is Your Team's Score?

WORKSHEET

Subject: Math
Grade: 1
Big Idea #1
Benchmark #1

____
Student Name

What Is Your Team's Score?

Students complete runs scored chart for each inning and add total.

	1	2	3	4	5	6	7	8	9	Total
Team										
Team										
Total Runs										

If a team did not score runs in the (name an inning), what would the total score have been?

	1	2	3	4	5	6	7	8	9	Total
Team	1	0	2	3	0	1	4	0	2	
Team	2	1	0	4	3	0	1	0	2	
Total Runs										

Which team had the most runs by the (name an inning)? Which team had scored less runs by the (name an inning)? Name each inning.

How many runs would need to be scored in the 4th inning to reach the total runs scored?

	1	2	3	4	5	6	7	8	9	Total
Team	0	1	3	__	2	0	1	3	1	14
Team	1	3	0	__	2	1	0	0	1	12
Total Runs										

LESSON PLAN

SUBJECT: MATH
GRADE: 1

<u>BIG IDEA #1</u>: Develop understandings of addition and subtraction strategies for basic addition facts and related subtraction facts.

<u>BENCHMARK #2</u>: (MA.1.A.1.2) Identify, describe, and apply addition and subtraction as inverse operations.

<u>TITLE OF LESSON</u>: Different Ways To Tie The Score In The Second Inning

<u>EQUIPMENT/MATERIALS</u>: Math counters: buttons, chips, cut-out baseballs, etc.

<u>PRE-DISCUSSION</u>: Each team can score the same number of runs by scoring a different number of runs in each inning.

<u>ACTIVITY</u>: Teacher has children complete the scores that total the number of runs on the worksheet. Students create scores in the empty inning boxes that equal the total runs scored.

<u>VARIATIONS</u>: Complete the missing number to equal the total number of runs scored.

<u>ASSESSMENT</u>: Student completes correct worksheet. Answers may vary and still be correct.

<u>WORKSHEET</u>: Add two (2) scores in two (2) innings to equal total number of runs.

WORKSHEET

Subject: Math
Grade: 1
Big Idea #1
Benchmark #2

____
Student Name

Different Ways To Tie The Score In The Second Inning

#1

Inning:	1		2		Runs
(example) Boston Red Sox	2	+	3	=	5
Tampa Bay	3	+	2	=	5

#2

Inning:	1		2		Runs
Baltimore	3	+	___	=	7
Philadelphia	4	+	___	=	7

#3

Inning:	1		2		Runs
Arizona	5	+	___	=	9
Minnesota	___	+	4	=	9

#4

Inning:	3		4		Runs
San Diego	3	+	___	=	11
Los Angeles	___	+	3	=	11

#5

Inning:	8		9		Runs
Pittsburgh	6	+	___	=	8
Arizona	___	+	6	=	8

LESSON PLAN

SUBJECT: MATH
GRADE: 1

BIG IDEA #1: Develop an understanding of addition and subtraction strategies for basic addition facts and related subtraction facts.

BENCHMARK #3: (MA.1.A.1.3) Create and use increasingly sophisticated strategies, and use properties such as Commutative, Associative and Additive Identity, to add whole numbers.

BENCHMARK #4: (MA.1.A.1.4) Use counting strategies, number patterns, and models as a means for solving basic addition and subtraction fact problems.

TITLE OF LESSON: What's The Score?

PRE-DISCUSSION: Discuss that an inning is after each team has had a turn to bat and that there are nine innings in a game.

ACTIVITY: Name the number of runs scored in each inning for the Tampa Bay Rays and Florida Marlins. Add the score after each inning and the total game.

VARIATIONS: Add the cost of tickets for a various number of people going to a game. Make up an estimated cost of food items and have students add the cost for each person.

ASSESSMENT: Students can correctly add the score of many different games.

WORKSHEET: What's The Score?

APPENDIX: (H) Ticket Costs

WORKSHEET

Subject: Math
Grade: 1
Big Idea #1
Benchmark #3,4

____
Student Name

What's The Score?

Inning	Tampa Bay Rays (runs scored)	Florida Marlins (runs scored)	Inning Score: Team	Inning Score	Inning Score: Team
1	0	2	TBR	2-0	FM
2	1	0	TBR	2-1	FM
3	1	0	TBR		FM
4	0	1	TBR		FM
5	2	1	TBR		FM
6	0	2	TBR		FM
7	1	0	TBR		FM
8	3	1	TBR		FM
9	0	0	TBR		FM
Total:	____	____	TBR		FM

VARIATIONS: Vary scores in other games on separate worksheets. Vary teams playing. Use number of hits, walks, pitches, strikeouts, etc. in a game.

LESSON PLAN

SUBJECT: MATH
GRADE: 1

BIG IDEA #2: Develop an understanding of whole number relationships, including grouping by tens and ones.

BENCHMARK #1: (MA.1.A.2.1) Compare and order whole numbers at least to 100.

BENCHMARK #2: (MA 1.A.2.2) Represent two digit numbers in terms of tens and ones.

TITLE OF LESSON: What's Your Uniform Number? Odd or Even

EQUIPMENT/MATERIALS: Base ten blocks, connecting blocks, or any 10 manipulatives

PRE-DISCUSSION: Make a group of 10 ones to represent 1, 10.

ACTIVITY: Discuss that each player has a number on their uniform. Have children cut out shirts on the worksheet and paste odd numbers on one sheet of colored paper and even numbers on the other.

VARIATIONS: Use batting averages in hundreds. Make up a final score and represent the number with manipulatives.

ASSESSMENT: Verbally or through observation. Complete a 10's and 1's chart correctly when given a number; 10, 1

WORKSHEET: Teacher provides 2 pieces of construction paper in two (2) different colors. Students cut out uniforms with various numbers and paste odd numbers on one sheet and even on the other. Students draw their own uniforms and use different numbers to group odd and even.

APPENDIX: (I) Tampa Bay Rays Player Averages

WORKSHEET

Subject: Math
Grade: 1
Big Idea #2
Benchmark #1 and 2

____
Student Name

What's Your Uniform Number? Odd or Even

LESSON PLAN

SUBJECT: MATH
GRADE: 1

BIG IDEA #2: Develop an understanding of whole numbers relationships, including grouping by tens and ones.

BENCHMARK #3: (MA.1.A.2.3) Order counting numbers, compare their relative magnitudes, and represent numbers on a number line.

TITLE OF LESSON: Display Your Teams Runs

EQUIPMENT/MATERIALS: Each child draws their own number line on a paper from the teacher's chart.

PRE-DISCUSSION: Concepts: before—after, greater than—less than.

ACTIVITY: Create a number line on the board or on a chart up to 10. Place the score of each team on the number line; Tampa Bay Rays—6, Florida Marlins—8; vary the scores and teams.

VARIATIONS: (1) Increase the number line and the number of runs scored. (2) Use the number of hits by 3 or 4 players on the number line.

ASSESSMENT: Child can complete missing numbers on a number line randomly.

LESSON PLAN

SUBJECT: MATH
GRADE: 1

BIG IDEA #3: Compose and decompose two-dimensional and three-dimensional geometric shapes.

BENCHMARK #1: (MA.1.G.3.1) Use appropriate vocabulary to compare shapes according to attributes and properties such as number and lengths of sides, and number of vertices.

BENCHMARK #2: (MA.1.G.3.2) Compose and decompose plane and solid figures, including making predictions about them, to build an understanding of part-whole relationships and properties of shapes.

TITLE OF LESSON: Find Baseball Shapes

EQUIPMENT/MATERIALS: Scissors, paper, base, baseball, home plate.

PRE-DISCUSSION: Shapes can be found in any environment. Can you name or find shapes at a baseball game?

ACTIVITY: Students discuss all shapes at a baseball park. Which are 2-dimensional and 3-dimensional? Which are straight lines or curved lines?

VARIATIONS: Describe properties of shapes and make predictions about them.

ASSESSMENT: Can students name, cut shapes into parts and re-assemble, and describe attributes and properties?

WORKSHEET: Cut out shapes. Cut shapes into parts and re-assemble as a puzzle. Name and label shapes.

WORKSHEET

Subject: Math
Grade: 1
Big Idea #3
Benchmark #1,2

____
Student Name

Find Baseball Shapes

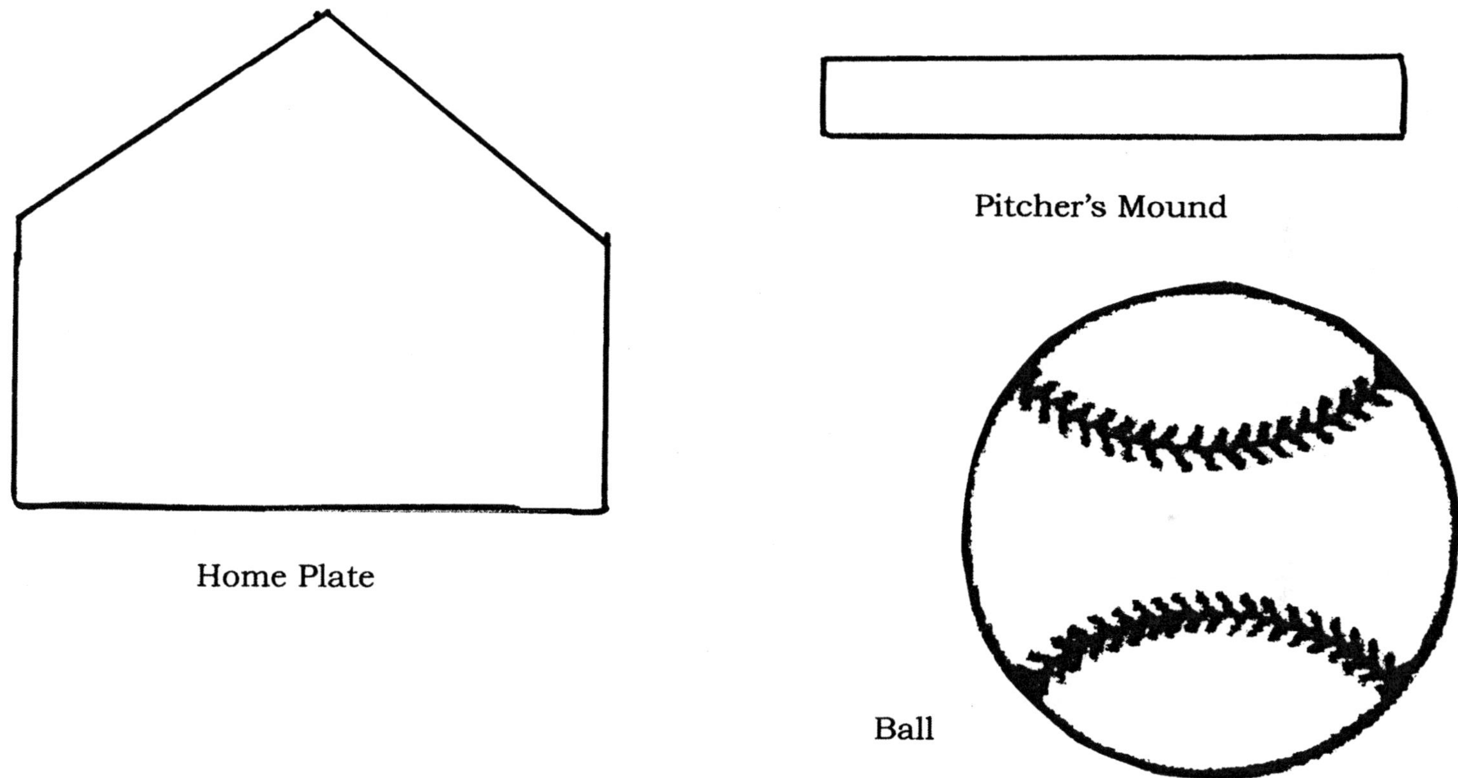

Cut out and cut in half. Re-assemble.

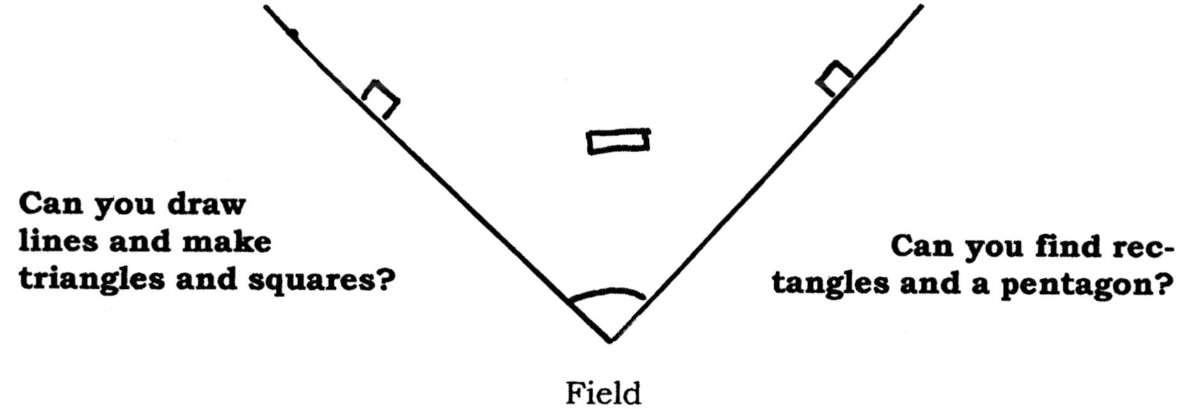

LESSON PLAN

SUBJECT: MATH
GRADE: 1

SUPPORTING IDEA: ALGEBRA

BENCHMARK #1: (MA.1.A.4.1) Extends repeating and growing patterns, fills in missing terms, and justifies reasoning.

TITLE OF LESSON: What's The Order?

EQUIPMENT/MATERIALS: Shoe, glove, bat, ball, base, hat, etc.

PRE-DISCUSSION: Discuss patterns in the environment.

ACTIVITY: Teacher provides baseball-related patterns and children repeat, complete and fill in missing item. Use various increasing numbers of items. Have children make up their own baseball patterns.

VARIATIONS: The function machine takes in 3 bats and puts out 6 bats. Then it takes in 2 gloves and puts out 8 gloves. What is the function? What is the rule?

ASSESSMENT: Have children make up patterns for baseball objects or knowledge.

WORKSHEET: Continue The Pattern.

WORKSHEET

Subject: Math
Grade: 1
Supporting Idea: Algebra
Benchmark #1

____
Student Name

Continue The Pattern

Horizontal Patterns:

Glove, Bat, Ball, Glove, ______________________, __________________________

Single, Double, Triple, Home Run, Single, ______________, ________________

____

1st Base, 2nd Base, 3rd Base, Home, 1st, ________________, ________________

____

Player, Coach, Umpire, Player, ____________________, _____________________

Tampa Bay Rays, New York Yankees, Philadelphia Phillies, Tampa Bay Rays,

__________________________________, ___________________________________

Seat Costs: $10.00, $20.00, $30.00, $10.00, _____________, ______________

Hot Dog, Pretzel, Soft Drink, Hot Dog, __________________, ________________

Vertical Patterns:

Glove	Player	Hot Dog
Bat	Coach	Pretzel
Ball	Umpire	Soft Drink
Glove	Player	Hot Dog
___________	___________	___________
___________	___________	___________

*Students draw baseball items to continue patterns.

LESSON PLAN

SUBJECT: MATH
GRADE: 1

SUPPORTING IDEA: ALGEBRA

BENCHMARK #2: Classifies numbers as odd or even and explain why.

TITLE OF LESSON: Odd Team, Even Team

EQUIPMENT/MATERIALS: Color paper, scissors.

PRE-DISCUSSION: How odd and even numbers are used in baseball.

ACTIVITY: Teacher provides cut-out numbers to students who trace, cut out and color their numbers to be placed on the back of student's shirts. Partners work together to tape numbers on their backs. Students group all of the odd number players on one team and even number players on the other team.

VARIATIONS: Use team runs scored, batting averages, attendance and identify whether numbers are odd or even.

ASSESSMENT: Place all of the odd-numbered players on one team and the even-numbered players on the other team.

WORKSHEET: Game—Students walk around the room and stop at someone to guess their number. They may give clues, i.e., between 1 and 9, odd or even, etc. See who is the closest to guess the correct number. Move to another student and repeat the process.

APPENDIX: (G) Attendance, (H) Ticket Costs and (M) Individual Statistics

LESSON PLAN

SUBJECT: MATH
GRADE: 1

SUPPORTING IDEA: GEOMETRY AND MEASUREMENT

BENCHMARK #1: (MA.1.G.5.1) Measures by using iterations of a unit and counts the unit measures by grouping units.

BENCHMARK #2: (MA.1.G.5.2) Compares and orders objects according to descriptors of length, weight and capacity.

TITLE OF LESSON: Baseball Measurements

EQUIPMENT/MATERIALS: Baseball bat and ball.

PRE-DISCUSSION: Use children's bodies laying down head to toe to estimate various distances; across the room, from one base to another, etc.

ACTIVITY: Students place a bat end over end to measure the number to get from one point to another. Repeat procedure with a student's shoe walking heel touching toe.

VARIATIONS: Move from concrete, a child's body, to semi-abstract, a baseball piece of equipment, to verbal, completing the worksheet.

ASSESSMENT: Students can successfully progress from estimation to concrete to academic.

WORKSHEET: Circle the word that best describes the baseball estimates.

APPENDIX: (C) Equipment

WORKSHEET

Subject: Math
Grade: 1
Supporting Idea: Geometry and Measurement
Benchmark #1 and 2

____
Student Name

Baseball Measurements

Circle the word that fits the best!

The bat is ______________ than the rest.
(long, longer, longest)

The ball is ______________ than the bat.
(light, lighter, lightest)

The player is the ______________ on his team.
(tall, taller, tallest)

The player is the ______________ on her team.
(heavy, heavier, heaviest)

The glove is ______________ than the ball.

A base is ______________ than a bat.

*Students make up others.

LESSON PLAN

SUBJECT: MATH
GRADE: 1

SUPPORTING IDEA: GEOMETRY AND MEASUREMENT

BENCHMARK #3: Identifies, combines and compares values of money.

TITLE OF LESSON: How Much Money Do You Need?

EQUIPMENT/MATERIALS: Discuss and show various dollar bills--$20, $10, $5 and $1.

PRE-DISCUSSION: Estimate the cost of various seats at a game. Why is there a difference?

ACTIVITY: Select your seat and cost. Add up the total cost for a family. Vary your seat selection, cost, and number of family members attending.

VARIATIONS: Identify a cost for various food items; hot dog, popcorn, soft drink, etc. Select which items each child wants and add cost. Make up a menu.

ASSESSMENT: Can students add the costs correctly?

APPENDIX: (H) Ticket Costs

LESSON PLAN

SUBJECT: MATH
GRADE: 1

SUPPORTING IDEA: NUMBERS AND OPERATIONS

BENCHMARK #1: (MA.1.A.6.1) Uses mathematical reasoning and beginning understanding of tens and ones, including the use of invented strategies, to solve two-digit addition and subtraction problems.

BENCHMARK #2: (MA.1.A.6.2) Solves routine and non-routine problems by acting them out, using manipulatives, and drawing diagrams.

TITLE OF LESSON: Play Math Baseball On A Field

EQUIPMENT/MATERIALS: 2-digit addition and subtraction cards. Four colored paper 3x5 cards with the teams initials on each; green Tampa Bay Rays, black New York Yankees.

PRE-DISCUSSION: Review 2-digit + and – facts using flash cards.

ACTIVITY: Take class to a baseball field. Divide class into 2 (two) teams. Show a student an addition flash card. If the correct answer is given, they may go to first base carrying their color team card. Each team gets an alternating turn to answer. Students advance to 1st, 2nd, 3rd and home to score runs. A member of each team can be on the same base.

VARIATIONS: Increase difficulty of problems. Vary baseball team names used.

ASSESSMENT: Observe and/or record the number of correct answers.

WORKSHEET: Keep score of the number of runs.

WORKSHEET

Subject: Math
Grade: 1
Supporting Idea: Numbers and Operations
Benchmark #1, #2

____
Student Name

Playing Math Baseball On A Field

Keep count of the number of runs scored by each team during the activity on this sheet. Re-use as repeat game, using different team names.

Tampa Bay Rays ____________________

Florida Marlins _________________________

- -

New York Yankees ______________________

Boston Red Sox ________________________

- -

Tampa Bay Rays ____________________

New York Yankees _______________________

SECOND GRADE

K-8 MATHEMATICS STANDARDS
GRADE 2

BIG IDEA 1: ***Develop an understanding of base-ten numerations system and place-value concepts.***	
BENCHMARK CODE	**BENCHMARK**
MA.2.A.1.1	Identify relationships between the digits and their place values through the thousands, including counting by tens and hundreds.
MA.2.A.1.2	Identify and name numbers through thousands in terms of place value and apply this knowledge to expanded notation.
MA.2.A.1.3	Compare and order multi-digit numbers through the thousands.

Access Points for Students with Significant Cognitive Disabilities

Independent:	*Supported:*	*Participatory:*
MA.2.A.1.In.a Apply the concept of grouping to create sets of tens and ones to 20 as a strategy to aid in counting. MA.2.A.1.In.b Represent numbers to 20 using sets of objects and pictures, number names, and numerals. MA.2.A.1.In.c Identify and use ordinal numbers to fifth. MA.2.A.1.In.d Use one-to-one correspondence to count, compare, and order whole numbers 0 to 20.	MA.2.A.1.Su.a Represent quantities to 5 or more using sets of objects, number names, and numerals. MA.2.A.1.Su.b Use one-to-one correspondence to count, compare, and order sets of objects to 5 or more.	MA.2.A.1.Pa.a Indicate desire to continue an action or activity by using an object in familiar routines. MA.2.A.1.Pa.b Respond to familiar actions that represent the concept of none or stop in routines. MA.2.A.1.Pa.c Respond to a counting cue to begin familiar routines in multiple settings. MA.2.A.1.Pa.d Match one object to a designated space to show one-to-one correspondence.

BIG IDEA 2: *Develop quick recall of addition facts and related subtraction facts and fluency with multi-digit addition and subtraction.*	
BENCHMARK CODE	**BENCHMARK**
MA.2.A.2.1	Recall basic addition and related subtraction facts.
MA.2.A.2.2	Add and subtract multi-digit whole numbers through three digits with fluency by using a variety of strategies, including invented and standard algorithms and explanations of those procedures.
MA.2.A.2.3	Estimate solutions to multi-digit addition and subtraction problems, through three digits.
MA.2.A.2.4	Solve addition and subtraction problems that involve measurement and geometry.

Access Points for Students with Significant Cognitive Disabilities

Independent:	***Supported:***	***Participatory:***
MA.2.A.2.In.a Identify the meaning of the +, -, and = signs in addition and subtraction problems. MA.2.A.2.In.b Use counting and one-to-one correspondence as strategies to solve problems involving addition facts with sums to 10 and related subtraction facts using numerals with sets of pictures. MA.2.A.2.In.c Solve real-world problems involving addition facts with sums to 10 and related subtraction facts, including money, measurement, geometry, and other problem situations.	MA.2.A.2.Su.a Identify the meaning of addition as adding to and subtraction as taking away from, using sets of objects. MA.2.A.2.Su.b Use counting and one-to-one correspondence as strategies to solve number stories involving addition facts with sums to 5 and related subtraction facts using sets of objects. MA.2.A.2.Su.c Solve real-world problems involving addition facts with sums to 5 and related subtraction facts using sets of objects.	MA.2.A.2.Pa.a Respond to the arrival of a familiar person or addition of a familiar object in the same activity in multiple settings. MA.2.A.2.Pa.b Respond to the departure of a familiar person or removal of a familiar object in the same activity in multiple settings. MA.2.A.2.Pa.c Continue in a routine with the addition of a familiar person, action, or object. MA.2.A.2.Pa.d Continue in a familiar routine with the removal of a familiar person, action, or object. MA.2.A.2.Pa.e Initiate a preferred action or activity by using an object.

BIG IDEA 3: *Develop an understanding of linear measurement and facility in measuring lengths.*	
BENCHMARK CODE	**BENCHMARK**
MA.2.G.3.1	Estimate and use standard units, including inches and centimeters, to partition and measure lengths of objects.
MA.2.G.3.2	Describe the inverse relationship between the size of a unit and number of units needed to measure a given object.
MA.2.G.3.3	Apply the Transitive Property when comparing lengths of objects.
MA.2.G.3.4	Estimate, select an appropriate tool, measure, and/or compute lengths to solve problems.

Access Points for Students with Significant Cognitive Disabilities		
Independent:	***Supported:***	***Participatory:***
MA.2.G.3.In.a Use standard units of whole inches to measure the length of objects. MA.2.G.3.In.b Compare and order objects of different lengths. MA.2.G.3.In.c Select and use a ruler to measure and compare lengths to solve problems.	MA.2.G.3.Su.a Measure the length of objects using nonstandard units of measure and count to 5 or more units. MA.2.G.3.Su.b Compare lengths of objects to solve real-world problems.	MA.2.G.3.Pa.a Respond to a prompt indicating size or length, such as big, little, long, or short in activities.

SUPPORTING IDEAS	
Algebra	
BENCHMARK CODE	**BENCHMARK**
MA.2.A.4.1	Extend number patterns to build a foundation for understanding multiples and factors – for example, skip counting by 2's, 5's, 10's.
MA.2.A.4.2	Classify numbers as odd or even and explain why.
MA.2.A.4.3	Generalize numeric and non-numeric patterns using words and tables.
MA.2.A.4.4	Describe and apply equality to solve problems, such as in balancing situations.
MA.2.A.4.5	Recognize and state rules for functions that use addition and subtraction.

Access Points for Students with Significant Cognitive Disabilities		
Independent:	***Supported:***	***Participatory:***
MA.2.A.4. In.a Identify two-element repeating visual patterns and extend with one repetition. MA.2.A.4.In.b Fill in missing items in two-element repeating visual patterns. MA.2.A.4.In.c Identify equal and unequal sets of objects and pictures to 20.	MA.2.A.4.Su.a Match two-element repeating patterns of sounds, physical movements, and objects. MA.2.A.4.Su.b Use one-to-one correspondence to identify sets of objects with the same number to 5.	MA.2.A.4.Pa.a Follow a two-element repeating pattern in a familiar routine or activity. MA.2.A.4.Pa.b Indicate anticipation of next step(s) in the same routine or activity in multiple settings. MA.2.A.4.Pa.c Recognize similarities and differences in features of familiar objects and actions in routines.

SUPPORTING IDEAS

Geometry and Measurement

BENCHMARK CODE	BENCHMARK
MA.2.G.5.1	Use geometric models to demonstrate the relationships between wholes and their parts as a foundation to fractions.
MA.2.G.5.2	Identify time to the nearest hour and half hour.
MA.2.G.5.3	Identify, combine, and compare values of money in cents up to $1 and in dollars up to $100, working with a single unit of currency.
MA.2.G.5.4	Measure weight/mass and capacity/volume of objects. Include the use of the appropriate unit of measure and their abbreviations including cups, pints, quarts, gallons, ounces (oz), pounds (lbs), grams (g), kilograms (kg), milliliters (mL) and liters (L).

Access Points for Students with Significant Cognitive Disabilities

Independent:	*Supported:*	*Participatory:*
MA.2.G.5.In.a Match parts with the whole using geometric shapes.	MA.2.G.5.Su.a Identify part and whole of geometric shapes.	MA.2.G.5.Pa.a Respond to differences in features such as size and shape of familiar objects in routines.
MA.2.G.5.In.b Identify concepts of time, including yesterday, today, tomorrow, first, and next, by relating activities with the time period.	MA.2.G.5.Su.b Match common three-dimensional objects by shape, including sphere and cube.	MA.2.G.5.Pa.b Respond to the environmental cue for a non-preferred activity within a regularly scheduled routine.
MA.2.G.5.In.c Identify the days of the week in relation to the calendar.	MA.2.G.5.Su.c Identify the concepts of time, including before, after, and next, by relating activities with the time period.	MA.2.G.5.Pa.c Respond to icons or symbols representing activities in an adaptive schedule.
MA.2.G.5.In.d Identify analog and digital clocks as tools for telling time.	MA.2.G.5.Su.d Identify coins as money.	MA.2.G.5.Pa.d Given a model, imitate one or more directional responses in a daily activity.
MA.2.G.5.In.e Identify the purpose of coins and bills.	MA.2.G.5.Su.e Compare weight of objects using the concepts of heavy and light.	
MA.2.G.5.In.f Compare objects by weight, using terms including heavy and light; and capacity, using terms including holds more and holds less.	MA.2.G.5.Su.f Identify and describe spatial relationships, including over, under, front, back, and between.	

SUPPORTING IDEAS		
Number and Operations		
BENCHMARK CODE	**BENCHMARK**	
MA.2.A.6.1	Solve problems that involve repeated addition.	
Access Points for Students with Significant Cognitive Disabilities		
Independent:	***Supported:***	***Participatory:***
MA.2.A.6.In.a Solve problems involving addition of the same number such as 1+1 or 2+2 with sums to 10.	MA.2.A.6.Su.a Solve problems involving combining sets with the same number of objects with sums to 4 using one-to-one correspondence and counting.	MA.2.A.6.Pa.a Communicate the desire for one preferred item or activity in familiar routines.

LESSON PLAN

SUBJECT: MATH
GRADE: 2

BIG IDEA #1: Develop an understanding of base-ten numerations system and place-value concepts.

BENCHMARK #1: (MA.2.A.1.1) Identifies relationships between the digits and their place values through the thousands, including counting by tens and hundreds.

BENCHMARK #2: (MA.2.A.1.2) Identifies and names numbers through thousands in terms of place value and applies this knowledge to expanded notation.

TITLE OF LESSON: Baseball Records

EQUIPMENT/MATERIALS:

PRE-DISCUSSION: Review Appendix O and discuss the Home Run records and an RBI (**R**un **B**atted **I**n).

ACTIVITY: Select a player and their total of home runs. Count by tens to reach their number. Count by hundreds to reach their number.

VARIATIONS: Use Runs Batted In records to count by tens and hundreds to reach the number. Select various players to repeat exercise.

ASSESSMENT: Have students write the numbers counting by tens and hundreds to the record number selected.

APPENDIX: (L) Major League Baseball Records, Home Runs and Runs Batted In

LESSON PLAN

SUBJECT: Math
GRADE: 2

BIG IDEA #1: Developing an understanding of base-ten numerations system and place-value concepts.

BENCHMARK #3: (MA.2.A.1.3) Compares and orders multi-digit numbers through the thousands.

TITLE OF LESSON: Who Is The Best Player?

PRE-DISCUSSION: Batting Averages—A batting average is the number of hits compared to the number of at-bat chances.

ACTIVITY: The students complete the worksheet by putting in order from top to bottom or best to lowest. Use single digits until understood, then increase to 2- and 3-digit.

VARIATIONS: Compare or order 2 or 3 numbers, then 3-5, then increase to level of understanding.

ASSESSMENT: The number of correct responses.

WORKSHEET: Who Is The Best Player?

APPENDIX: Tampa Bay Rays Individual Statistics (J)

WORKSHEET

Subject: Math
Grade: 2
Big Idea #1
Benchmark #3

____
Student Name

Who Is The Best Player?

(H) HITS	ORDER HIGHEST TO LOWEST	(G) GAMES PLAYED	ORDER LOWEST TO HIGHEST
60		30	
83		21	
87		42	
55		58	
23		56	
23		114	
		21	

(BA) BATTING AVERAGE	NAME ORDER	(E) ERRORS	NAME ORDER
317		0	
305		3	
260		1	
244		5	
224		8	
228		13	
		7	
		4	

LESSON PLAN

SUBJECT: MATH
GRADE: 2

BIG IDEA #2: Develop quick recall of addition facts and related subtraction facts and fluency with multi-digit addition and subtraction.

BENCHMARK #1: (MA.2.A.2.1) Recalls basic addition and related subtraction facts.
BENCHMARK #2: (MA.2.A.2.2) Adds and subtracts multi-digit whole numbers by using a variety of strategies.
BENCHMARK #3: (MA.2.A.2.3) Estimates solutions to multi-digit addition and subtraction problems, through three digits.
BENCHMARK #4: (MA.2.A.2.4) Solves addition and subtraction problems that involve measurement and geometry.

TITLE OF LESSON: Math Baseball Game

EQUIPMENT/MATERIALS: Questions and flash cards that:

1) recall basic + and - facts,
2) + and - multi-digit whole numbers,
3) when to use operations and why they work, and
4) estimates multi-digit + and - problems.

PRE-DISCUSSION: Select 2 Major League Baseball teams and identify their colors. Make 6 (3x5) cards with their team colors.

ACTIVITY: Students at a baseball field on two different MLB teams are asked similarly difficult questions. Alternate questions to students from both teams. If the correct answer is given, the student goes to first base. Students move one base at a time. Collect the 3x5 cards as the students come home and score a run.

ASSESSMENT: Record the number of correct answers for each student.

LESSON PLAN

SUBJECT: MATH
GRADE: 2

BIG IDEA #3: Develops an understanding of linear measurement and facility in measuring lengths.

BENCHMARK #1: (MA.2.G.3.1) Estimates and uses standard units of linear measurement, both centimeter and inch, to partition and measure lengths of objects.
BENCHMARK #4: (MA.2.G.3.4) Estimate, select an appropriate tool, measure, and/or compute lengths to solve problems.

TITLE OF LESSON: Select Your Team At The Tryouts

EQUIPMENT/MATERIALS: Stopwatch, scale, tape measure, calculator.

PRE-DISCUSSION: What could baseball players do to show the coach that they are the best? How could you measure how good they are?

ACTIVITY: The teacher presents skills that players need to demonstrate to be very good players. Students select the best instrument to measure their performance. Student can draw or write the answer in the space on the worksheet.

VARIATIONS: Create other measuring devices to measure performance. Review the batting averages of the Tampa Bay Rays players. Select a team of your classmates.

ASSESSMENT: Students choose the correct measuring device for each situation.

WORKSHEET: Select Your Team!

APPENDIX: (I) Tampa Bay Rays Statistics

WORKSHEET

Subject: Math
Grade: 2
Big Idea #3
Benchmark #1, #4

____
Student Name

SELECT YOUR TEAM AT THE TRYOUTS!

Object used to measure performance:

The players throw a ball as far as they can __________________________

The players run a 50-yard dash ____________________________________

The players take turns batting: how many times did they hit the

ball?____

The players run around the bases __________________________________

Students select other skills and choose the object to measure performance.

LESSON PLAN

SUBJECT: MATH
GRADE: 2

BIG IDEA #3: Develops an understanding of linear measurement and facility in measuring lengths.

BENCHMARK #2: (MA.2.G.3.2) Describes the inverse relationship between size of unit and number of units needed to measure a given object.

TITLE OF LESSON: Baseball Measurements

EQUIPMENT/MATERIALS: One-foot ruler and yardstick, baseball shoe, baseball bat.

PRE-DISCUSSION: Review the size of these objects and determine which is greater than and less than the other.

ACTIVITY: Using the distance from home plate to first base as the goal, use the one-foot ruler, shoe, bat, their step and their body length and have the students respond to greater than and less than.

VARIATIONS: Visit the baseball field and do actual measurements of equipment from base to base. Vary measurements using centimeters and meters.

ASSESSMENT: Correct estimations.

WORKSHEET: Baseball Measurements

WORKSHEET

Subject: Math
Grade: 2
Big Idea #3
Benchmark #3

____
Student Name

Baseball Measurements

Questions:

1) Will it take more shoes or bats to go from home plate to 1st base?

2) Will it take more running steps or body lengths to go from 2nd base to 3rd base?

3) Will it take more or less bats to go from 1st base to 2nd base than measured with body lengths?

4) Will it take more yardsticks or one- foot rulers to go from 2nd base to 3rd base?

Have students make up more-than or less-than situations using the measurement devices specified or others (base etc.).

LESSON PLAN

SUBJECT: MATH
GRADE: 2

BIG IDEA #3: Develops an understanding of linear measurement and facility in measuring lengths.

BENCHMARK #3: (MA.2.G.3.3) Applies the Transitivity Property when comparing lengths of objects.

TITLE OF LESSON: Graphing Baseball Equipment

EQUIPMENT/MATERIALS: A baseball, baseball shoe, base, and a bat.

PRE-DISCUSSION: Establish that the order of length of the following is: a baseball, baseball shoe, a base, a student's step, a bat, the distance from home to 1st base, etc.

ACTIVITY: Select any 3 of the above measurements in the pre-discussion and apply the formula that if A is greater than B and B is greater than C, then A is greater than C.

VARIATIONS: Have students apply the formula to other measurements in baseball.

ASSESSMENT: Correct responses. Assess through comparison questions—which is biggest, smallest, etc.

LESSON PLAN

SUBJECT: MATH
GRADE: 2

SUPPORTING IDEA: GEOMETRY AND MEASUREMENT

BENCHMARK #1: (MA.2.G.5.1) Uses geometric models to demonstrate the relationships between wholes and their parts as a foundation to fractions.

TITLE OF LESSON: Baseball Puzzles

EQUIPMENT/MATERIALS: Pictures of baseballs, shoes, gloves, bats, bases, fields, etc.

PRE-DISCUSSION: Parts put together equal a whole.

ACTIVITY: Students draw pictures of materials listed above and cut them into various numbers of pieces and re-assemble as a puzzle.

VARIATIONS: Duplicate Appendix Y—Stadium Seating Chart. Students cut the diagram into various pieces and re-assemble as a puzzle.

ASSESSMENT: Puzzle completion

WORKSHEET: Stadium Seating Chart

APPENDIX: (Y) Stadium Seating Chart

LESSON PLAN

SUBJECT: MATH
GRADE: 2

SUPPORTING IDEA: ALGEBRA

BENCHMARK #1: (MA.2.A.4.1) Extends number patterns to build a foundation for understanding multiples and factors—for example, skip counting by 2's, 5's, 10's.

BENCHMARK #2: (MA.2.A.4.2) Classifies numbers as odd or even and explains why.

BENCHMARK #3: (MA.2.A.4.3) Generalizes numeric and non-numeric patterns using words and tables.

TITLE OF LESSON: Runs Scored

EQUIPMENT/MATERIALS: Newspaper—Sports Section

PRE-DISCUSSION: Look in a newspaper to obtain examples of runs scored by various teams.

ACTIVITY: Have students identify an amount of runs scored in a game and count to that number by twos, threes, fives, etc.

VARIATIONS: Use non-numeric patterns such as ball, bat, baseball, ____, ____, ____, ____, ____.

ASSESSMENT: Correct responses

WORKSHEET: Runs Scored. Have students complete the inning chart using two runs scored in each, three, four, etc. Vary the pattern and then have students create their own pattern.

APPENDIX: (S) Seating Chart. Reproduce and white out every other section or every third section. Have students complete correct section numbers.

WORKSHEET

Subject: Math
Grade: 2
Supporting Idea: Algebra
Benchmark #1,2 and 3

____
Student Name

Runs Scored

INNING	1	2	3	4	5	6	7	8	9
RUNS	2	4	6	___	___	___	___	___	___

INNING	1	2	3	4	5	6	7	8	9
RUNS	3	6	9	___	___	___	___	___	___

INNING	1	2	3	4	5	6	7	8	9
RUNS	4	8	___	___	___	___	___	___	___

INNING	1	2	3	4	5	6	7	8	9
RUNS	1	1	2	2	3	___	___	___	___

INNING	1	2	3	4	5	6	7	8	9
RUNS	1	2	3	1	2	3	___	___	___

INNING	1	2	3	4	5	6	7	8	9
RUNS	___	___	___	___	___	___	___	___	___

LESSON PLAN

SUBJECT: MATH
GRADE: 2

SUPPORTING IDEA: ALGEBRA

BENCHMARK #4: (MA.2.A.4.4) Describes and applies equality to solve problems, such as in balancing situations.

TITLE OF LESSON: What Does Your Baseball Equipment Weigh?

EQUIPMENT/MATERIALS: Draw a scale that is balanced on a large paper or board.

PRE-DISCUSSION: Different numbers of objects are needed to balance a scale if their weights are different.

ACTIVITY: Students complete the weights of various objects on the worksheet to balance on the scale. Create a weight equation (see examples on worksheet--2 baseballs equal the weight of a glove; if the glove equals 10 ounces, how many ounces does each baseball weigh?)

VARIATIONS: Vary the number of objects and their weights to balance the scale.

ASSESSMENT: Correct responses. Use a chart for comparisons.

WORKSHEET: Balance The Scale

WORKSHEET

Subject: Math
Grade: 2
Supporting Idea: Algebra
Benchmark #4

____
Student Name

Balance The Scale

(ball)(ball)
? + ? = glove
ounces = 10 ounces

(ball) (ball) (ball)
? + ? + ? = glove
ou. ou. ou. = 15 ou.

(glove) (glove)
? + ? = bat
ounces ounces = 16 ounces

shoe glove
? = ?
ou. ou.

Students create the weights of various baseball equipment items to balance the scale.

LESSON PLAN

SUBJECT: MATH
GRADE: 2

SUPPORTING IDEA: ALGEBRA

BENCHMARK #5: (MA.2.A.4.5) Recognizes and states rules for functions that use addition and subtraction.

TITLE OF LESSON: Batting Averages

EQUIPMENT/MATERIALS: None.

PRE-DISCUSSION: Batting averages are determined by number of hits versus number of at-bats. The player's averages go up and down.

ACTIVITY: Use an example of applying a + or – to a batting average (example: .306 - .294). Students complete worksheet.

VARIATIONS: Use the number of hits a player gets from one game to another.

ASSESSMENT: Number of correct responses.

WORKSHEET: Batting Averages

APPENDIX: (J) Tampa Bay Rays Individual Statistics

WORKSHEET

Subject: Math
Grade: 2
Supporting Idea: Algebra
Benchmark #5

____
Student Name

Batting Averages

Fill in a
+ or -
to show how
average changed

.196 ____ .204

.312 ____ .298

.265 ____ .273

.316 ____ .312

.243 ____ .232

.283 ____ .278

.241 ____ .253

.259 ____ .264

.281 ____ .279

.248 ____ .261

.319 ____ .309

.319 ____ .331

.296 ____ .304

.281 ____ .273

.324 ____ .319

.308 ____ .318

Rule
Place word
"Addition" or "Subtraction"
on line

.242 ______________ .164

.314 ______________ .309

Students create their own problem:

Average	Function	Average
.____	______________	.____
.____	______________	.____
.____	______________	.____
.____	______________	.____
.____	______________	.____
.____	______________	.____
.____	______________	.____
.____	______________	.____
.____	______________	.____
.____	______________	.____
.____	______________	.____
.____	______________	.____

LESSON PLAN

SUBJECT: MATH
GRADE: 2

SUPPORTING IDEA: GEOMETRY AND MEASUREMENT

BENCHMARK #1: (MA.2.G.5.1) Uses geometric models to demonstrate the relationships between wholes and their parts as a foundation to fractions.

TITLE OF LESSON: Baseball Puzzles

EQUIPMENT/MATERIALS: Collect pictures of baseballs, shoes, gloves, bats, bases, fields, etc. from newspapers and magazines.

PRE-DISCUSSION: Parts put together equal a whole.

ACTIVITY: Students draw pictures of materials listed above and cut them into various numbers of pieces and re-assemble as a puzzle.

VARIATIONS: Duplicate Appendix Y—Stadium Seating Chart. Students cut the diagram into various pieces and re-assemble as a puzzle. Adjust the number of pieces to students' ability. Ten to fifteen is recommended.

ASSESSMENT: Puzzle completion

WORKSHEET: Stadium Seating Chart

APPENDIX: (S) Stadium Seating Chart

LESSON PLAN

SUBJECT: MATH
GRADE: 2

SUPPORTING IDEA: GEOMETRY AND MEASUREMENT

BENCHMARK #2: (MA.2.G.5.2) Identifies time to the nearest hour and half-hour.

TITLE OF LESSON: Play Ball!

EQUIPMENT/MATERIALS: Clock, newspaper sports section, TV schedule

PRE-DISCUSSION: If students play baseball or softball, what time do their games begin? Discuss hour and half hour concept.

ACTIVITY: Establish a game time to "Play Ball" and have students identify the nearest half-hour or hour.

VARIATIONS: Select the nearest half-hour and hour to various game times. If a game starts at 7:15 and ends three (3) hours later, name the time. If a game starts at 1:30 and ends two-and-a-half (2.5) hours later, name the time.

ASSESSMENT: Correct responses.

LESSON PLAN

SUBJECT: MATH
GRADE: 2

SUPPORTING IDEA: GEOMETRY AND MEASUREMENT

BENCHMARK #3: (MA.2.G.5.3) Identify, combine and compare values in cents up to $1 and in dollars up to $100, working with a single unit of currency.

TITLE OF LESSON: How Much Does It Cost?

EQUIPMENT/MATERIALS: Play money, coins and bills.

PRE-DISCUSSION: Arrive at an agreed-upon cost of a hot dog, soft drink, popcorn and cotton candy.

ACTIVITY: Children are given coins in various amounts and display the amount to purchase the food items at determined costs. Combine a number of food items to arrive at different costs.

VARIATIONS: Discuss the cost of tickets for different seats. Using play money, show the correct amount to purchase one or more tickets at various prices. Providing change also develops concepts.

ASSESSMENT: Children show correct amounts of coins to buy food and dollars to buy tickets.

APPENDIX: (S) Stadium Seating Chart; (H) Tickets Costs

LESSON PLAN

SUBJECT: MATH
GRADE: 2

SUPPORTING IDEA: GEOMETRY AND MEASUREMENT

BENCHMARK #4: (MA.2.G.5.4) Measure weight/mass and capacity/volume of objects.

TITLE OF LESSON: You Are The Team Equipment Manager

EQUIPMENT/MATERIALS: Baseballs, gloves, shoes, bats, scale.

PRE-DISCUSSION: Estimate the weight of each equipment item. Estimate the capacity of a bag to hold balls, gloves, etc. Record each child's guess and see who is the closest to the correct answer.

ACTIVITY: Weigh various equipment items using as many different measurements as possible. Fill various size bags with equipment items.

VARIATIONS: What items would each player need for a game? How many items would a team need for a game?

ASSESSMENT: Children can measure the weight of equipment and volume of containers accurately.

APPENDIX: (C) Baseball Equipment; (K) Bat Size

LESSON PLAN

SUBJECT: MATH
GRADE: 2

SUPPORTING IDEA: NUMBER AND OPERATIONS

BENCHMARK #1: (MA.2.A.6.1) Solves problems that involve repeated addition.

TITLE OF LESSON: Working At A Baseball Stadium

EQUIPMENT/MATERIALS: None.

PRE-DISCUSSION: Establish the cost of a ticket using Appendix H. Establish how much money would be made a day selling various items like soda, hot dogs, cotton candy, etc. How much would you need to sell to purchase a ticket? Two or three tickets?

ACTIVITY: Create a story of a student wanting to go to a ballgame and purchase a ticket. Establish how much money is made in a day working and selling various items, and how many days it would take working to earn enough money to buy a ticket.

VARIATIONS: How many days would it take to work to purchase 2, 3, 4, etc. tickets?

ASSESSMENT: Correct calculations.

APPENDIX: (H) Ticket Costs

WORKSHEET

Subject: Math
Grade: 2
Supporting Idea:
Numbers and Operations
Benchmark #1

____
Student Name

Working At A Ballpark

Item
Candy

#1 #2	$2.00 $4.00
Ticket Cost	$12.00

Item
Hot dog

#1 #2	$3.00 $6.00
Ticket Cost	$18.00

Item
Soda

#1 #2	$4.00
Ticket Cost	$36.00

Other Items

APPENDIX

APPENDIX

A. BASEBALL HISTORY

1. A BRIEF BASEBALL HISTORY
2. BASEBALL ORIGINS

B. BASEBALL FIELD AND POSITIONS

C. BASEBALL EQUIPMENT

D. MINOR LEAGUE TEAMS

E. MAJOR LEAGUE TEAMS

F. FINAL STANDINGS 2007

G. ATTENDANCE

H. TICKET COSTS

I. TAMPA BAY RAYS TEAM AND SEASON RECORDS

J. TAMPA BAY RAYS INDIVIDUAL STATISTICS

K. BASEBALL BAT SIZES

L. MAJOR LEAGUE BASEBALL ALL-TIME RECORDS

M. "WHO'S ON FIRST"

N. CASEY AT THE BAT

O. BASEBALL AT 9-11-2001

P. BASEBALL DEFINITIONS

Q. BASEBALL BOOKS

R. BASEBALL-RELATED WEBSITES

S. STADIUM SEATING CHART—TAMPA BAY RAYS

---

www.majorleaguebaseball.com, Feb. '07

APPENDIX A

A BRIEF BASEBALL HISTORY

Early History

Baseball developed from variations of the English game of rounders, from related regional and local games, and from children's games like "one old cat," all of which had evolved through centuries. In the 1840's Alexander Cartwright of the New York Knickerbocker Club standardized many of the features and field dimensions still in use today. Sportswriter Henry Chadwick wrote (1858) the first rule book, and though the rules continue to change by small degrees, by 1900 the game was essentially that of today. The story that Abner Doubleday invented baseball in 1839 has been discredited.

The Development of Professional Baseball

In the mid-19th century baseball was primarily popular among local clubs in the Northeast, often made up of members of the same occupation. Eventually competition broadened, and an organization to promote standardized rules and facilitate scheduling, the National Association of Baseball Players, was formed in 1858. The movement of Union soldiers during the Civil War helped to spread the game, and increased opportunities for leisure, improved communications, and easier travel after the war fostered a wider competitive base and increased interest.

In 1869, Harry Wright organized the Cincinnati Red Stockings, baseball's first professional team, and took them on a 57-game national tour, during which they were unbeaten. Seeking to expand on the Reds' success, the National Association of Professional Baseball Players in 1871 chartered nine teams in eight cities as the first professional league. In the 1870's a number of competing leagues were formed, including the National League, which soon became the predominant association.

Financial hardships, gambling-related scandals, and franchise upheaval plagued all the leagues, and a players' revolt in 1890, which resulted in a short-lived Players Association, weakened the National League. A competing league, the Western Association, changed its name to the American League in 1900 and placed clubs in several eastern cities. In 1903 the champions of the American and National Leagues met for the first time in what became known as the World Series.

Both leagues fought off the challenge of the Federal League in 1914-15, but baseball's popularity and stability were threatened when the 1919 Chicago White Sox conspired to lose the World Series. Club owners then hired Judge Kenesaw Mountain Landis as the first baseball commissioner (1920-44) and charge him with resolving the crisis. Landis banned eight members of the "Black Sox" for life (despite their acquittal in a court of law), helping to lift suspicion from the professional game.

The Golden Years

The years between 1920 and World War II were the heyday of Babe Ruth, the game's preeminent legend. Other stars made their names as well: Ruth's durable New York Yankees teammate, Lou Gerhig; the contentious bating champion Ty Cobb; outstanding pitchers like Lefty Grove, Dizzy Dean, and Walter Johnson; graceful Yankee center fielder Joe DiMaggio; and sluggers Hank Greenberg and Jimmie Foxx, among others. Fans flocked to the large stadiums built in the 1920's.

Integration of Professional Baseball

During World War II, many major league stars served in the armed forces. By the mid-1940's, most had returned to their teams, but major league baseball continued to exclude black players, who, barred by a color line drawn in the 1880's, showcased their skills in separate leagues, especially the Negro National League (1920) and the Eastern Colored Leagues (1923). Black players like Satchel Paige, Buck Leonard, josh Gibson, and Judy Johnson, among the best in baseball, often played before large crowds, "invisible" to the white public. In 1947, Branch Rickey, Brooklyn's general manager, began the integration of the major leagues by bringing Jackie Robinson to the Dodgers. Weathering the great pressure and the hatred of many players and fans, Robinson became one of the most electrifying performers in the game, paving the way for other black stars like Willie Mays and Hank Aaron. *(continued on next page)*

Expansion and Labor Conflict

The locations of major league franchises, stable for 50 years, became unsettled in the 1950's. The Boston Braves moved to Milwaukee in 1953, and other teams joined a westward migration made feasible by the expansion of air travel and attractive by population shifts (and, ultimately, by the promise of regional television coverage). The 1957 exodus of the Brooklyn Dodgers and the New York Giants for California jarred New Yorkers but helped cement the game's nationwide base. In 1961, the two major leagues entered into a period of expansion, gradually adding new teams.

In the 1960's and 1970's, however, baseball's popularity was challenged by disillusionment of the young with established institutions, by the television-spurred boom of the national Football League (television was also presumed largely responsible for the shrinkage of the minor-league system), and by divisiveness within the sport over new artificial playing surfaces, indoor stadiums, and rule changes like the American League's 1973 introduction of a designated hitter to bat for the pitcher (the National League never adopted the measure).

Player-club relations were tumultuous in the 1970's. The Major League Baseball Players' Association, formed in 1966, pushed for an end to the reserve clause, a contractual stipulation that bound a player to one club unless he was traded, released, or retired. Although the U.S. Supreme Court had upheld the clause three times in 50 years, a mid-1970's arbitrator declared several players "free agents," and thereafter the sport was obliged to allow freer player movement among bidding teams. The Players' Association continued through the 1970's and the 1980's to strengthen the bargaining positions, salaries and pensions of the players. Conflict between team owners and players resulted in numerous work stoppages after 1972, the worst of which canceled the final third of the 1994 season, including the World Series.

Despite these distractions, however, the major-league game continued to flourish. As Babe Ruth was held to have carried the game through the post-Black Sox era, the breaking of Lou Gerhig's consecutive games played record in 1996 by Cal Ripken, Jr. and the assult on the single season home run record by Mark McGwire and Sammy Sosa in 1998 was seen as "rescuing" the game from its self-inflicted troubles. By the late 1990's there were 30 teams in six divisions in the major leagues (limited interleague play was introduced in 1997), attendance and television revenues were high, and talk about eventual expansion into Latin America and Asia was heard.

Bibliography

See L.S. Ritter, *The Glory of their Times* (1966); D. Voight, *American Baseball* (3 Vol., 1966-83); R.W. Peterson, *Only the Ball was White* (1970); H. Seymour, *Baseball:The Early Years* (1960), *Baseball: The Golden Age* (1971), *Baseball: The People's Game* (1990); G. Ward and K. Burns, *Baseball* (1994); *Baseball Encyclopedia* (10th ed. 1996); J. Thorn et al., *Total Baseball* (6th ed. 1999).

A. BASEBALL HISTORY

The exact origins of baseball are unclear, but it is thought that it developed from the English game of Rounders, which was brought to America in the early 1600s by the first English settlers. Different forms of the game were played in different parts of America, but all involved hitting a pitched ball with a stick or bat and trying to score runs. In 1846, Alexander Cartwright wrote a set of rules that form the basis of modern baseball. Today, baseball is the classic All-American sport and is played in over 100 countries around the world.

The first game to be played under the Cartwright Rules was held in 1846 at the Elysian Fields in New Jersey. The Knickerbockers beat the New York Nines by 23-1 and Alexander Cartwright, the writer of the rules, umpired the game.

Adrian "Cap" Anson was perhaps the most influential player in the 19th century. He played in the first professional league, the National Association, and was the first man to reach 3,000 hits.

One of the most important events in modern baseball occurred in 1947. Until that date, baseball had been a segregated sport, with separate leagues for black players. However, during that year, Jackie Robinson opened the season at second base for the Brooklyn dodgers. Robinson was the first black player to play in the Major Leagues.

The first professional baseball club, the Cincinnati Red Stockings, was formed in 1869. The club's manager was Harry Wright of Sheffield, England. The Red Stockings toured the U.S., playing against local teams.

Women have never played in any Major League baseball games, but there have been women's professional baseball teams, most notably during WWII. This league was made famous by the movie "A League of Their Own," which told the story of the All-American Girls' Professional Baseball League from 1943 to 1954.

Amateur baseball is played throughout the world, with slight variations. In a picture found of a game played in Montreal, Canada, in the 1952s, for example, the umpire is wearing his chest protector outside of his shirt.

Babe Ruth was one of the most famous players in the history of baseball. He joined the New York Yankees in 1920. During his 22-year career, the powerful left0-hander hit 714 home runs – a record until Henry Aaron broke it in 1974!

Since 1992, baseball has been a full-medal sport in the Olympic Games. In recent years, Cuba has dominated amateur and Olympic baseball. In 1996, the final was between Cuba and Japan. At the 2000 Olympic Games in Sydney, Australia, the U.S. won the gold medal for the first time by defeating Cuba 4-1. The Cuban team won the silver medal and the South Korean team took the bronze.

Once you have mastered the basic techniques involved in playing baseball, you may decide that you want to take the sport further. One of the biggest sports programs in the world is Little League Baseball, in which girls and boys up to the age of 18 years can participate. Every year, at Williamsport, Pennsylvania, eight teams compete to win the Little League

World Series. This series is for players aged 11 and 12, and attracts the best teams from all over the world. From here, you may want to progress to amateur baseball or even to the professional Minor and Major Leagues.

The prestigious World Series has been played since 1905. Today, the winning teams from the Eastern and Western Divisions of Major League teams meet to battle it out for the coveted trophy.

First American Edition, 1998
Published in the United States by Dorling Kindersley Publishing, Inc.
95 Madison Avenue, New York, NY 10016

B. BASEBALL PLAYING FIELD, POSITIONS

A baseball field is made up of the infield, the outfield, and foul territory. Two "foul lines" run from home plate through first and third bases to create a 90 degree arc. The infield is the inner part of the arc. A base is placed on each corner of the infield, 90 feet (28 meters) apart, to form the diamond. The outfield is the outer part of the baseball field. It is formed by the two foul lines extending to an outfield fence, which should be at least 250 feet (76 meters) from home plate. The area inside the foul lines is fair territory; the area outside is foul territory.

Field Positions:

Position	Abbreviation	Number
Pitcher	p	1
Catcher	c	2
First base	1b	3
Second base	2b	4

Third base	3b	5
Shortstop	ss	6
Left field	lf	7
Center field	cf	8
Right field	rf	9

The Game:

Winning a game. The object of the game is to score more runs than the opposition. To score a run, the batter/runner must touch each base in turn, and return to home plate.

Officials. The rules of the game are enforced by a crew of umpires. In the major leagues, there are four umpires per game. In youth baseball, there are usually two – one home plate umpire and one base umpire.

Game duration. An inning is completed when each team has had a turn at batting and at fielding. A standard baseball game lasts for nine innings. Each team has three outs per inning. The number of innings may be reduced in youth baseball.

Extra innings. A baseball game cannot end in a tie. One team has to win and extra innings must be played to ensure this result.

Offense and defense. The team that is batting is said to be playing offense, and the team in the field is the defending team.

Batting order. Before the game begins, the manager of each team decides on the batting order. This is written down and handed to the umpire. It cannot be changed during the game.

Substitutes. Substitutes can be used in a baseball game. However, once a player has been taken out of a game, he cannot return.

The playing field. The defensive positions on the baseball field are shown on the diagram. The pitcher, the catcher, and the four infielders defend the infield and the three outfielders defend the outfield.

The lf fields deep,
protecting the left field foul line

The ss and 3b defend
the left side of the diamond

The p stands on
the mound to pitch

The batting team waits in
the bench area, which is
also called the dugout

The cf fields deep,
protecting center field

The rf fields deep,
protecting the right field foul line.

1b and 2b defend
the right side of the diamond

The c's box is behind
home plate so that the
c is in a position to
receive the ball from
the pitcher

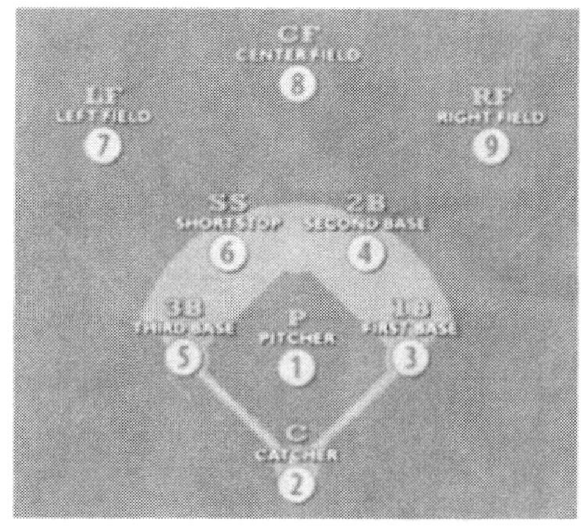

Base. There are three bases, one in each corner of the diamond. Each base is filled with foam and covered with white canvas. It measures 15 x 15 inches (38 x 38 centimeters) and is secured to the ground. When running the bases, you must touch the base, or you may be called out.

Pitcher's mound. The pitcher pitches the ball from a mound that is 10 inches (25 centimeters) above the level of the home plate area, and has a diameter of 18 feet (5.5 meters). There is a pitching rubber on top of the mound that is 60.5 feet (18.4 meters) from the back of home plate. The pitcher must in contact with the rubber when he starts his delivery.

The batter's box. You must be in the batter's box when you hit the ball. Assume your batting stance, and the strike zone will be the area from your chest to the bottom of your knees. The umpire calls a strike when the batter does not swing at a pitch in the strike zone, or swings at a pitch and misses, or hits the ball into foul territory with less than two strikes against him or her. A ball is a pitch delivered outside the strike zone that the batter does not swing at. After four balls, the batter can walk to first base.

First American Edition, 1998
Published in the United States by Dorling Kindersley Publishing, Inc.
95 Madison Avenue, New York, NY 10016

C. BASEBALL EQUIPMENT

All you really need to play baseball is a bat, a ball, and a glove. However, to get the most enjoyment from the game, you should also have all the necessary safety equipment. Baseball is a hard-ball game, and protective equipment for the catcher, batter, and base runners is mandatory. When playing in a league, each team has its own colors and logo and players wear identical uniforms, making them look neat and professional.

Baseball uniform. Players on baseball teams often wear a full baseball uniform. This consists of a cap, which is usually embroidered with the team logo, a baseball shirt with the team name and the player's number on it, an undershirt, baseball pants and a pair of stirrups that are worn over athletic socks.

Undershirt. Players may wear an undershirt under their team shirt. Long-sleeved and made of cotton, it absorbs perspiration.

Stirrups. Stirrups are unique to baseball. They are pull-over socks that are worn over athletic socks and are colored to match team uniforms.

Cleats. Cleats are similar to soccer shoes, except they have cleats or bars on the sole instead of studs. They help the players grip the ground securely.

Catcher's equipment. A catcher can receive over 100 fast pitches in a game and must have special equipment to protect him from injury. This includes a helmet, mask, chest protector, cup, and leg guards. Modern equipment gives good protection while still allowing the catcher to move around freely.

Leg guards are held in place with straps and clip buckles. They can be quickly and easily removed when the catcher is at bat. The face mask provides good protection and full vision. The mask is padded to soften the impact of a misplaced ball. The chest protector shields the upper body. A catcher's mitt may be larger than a normal fielder's glove, and has extra padding. Leg guards are hinged to allow a wider range of movement. Leg guards also protect the shins. Some guards come down over the shoes to protect your feet.

Basic equipment. There are several essential items that you will need to play baseball successfully and safely.

A catcher's helmet is mandatory in baseball, and will protect you from swinging bats and balls.

A batting helmet must be worn by the batter and any base runners. The helmet has a plastic shell with foam inserts. It protects the head and face from a thrown or batted ball.

The catcher's mitt has extra padding to protect his hand. Fielder's gloves are made of leather and come in a variety of sizes. Pitchers and outfielders use large gloves; infielders use smaller gloves.

A batting glove is worn to get a really firm grip on the bat. Players can wear two or just one on their bottom hand.

Bats come in various lengths and weights and can be made from either aluminum or wood. When choosing a bat, remember that a lighter bat is easier to swing.

The baseball itself is made of a cork core wrapped with string and covered in leather. It weighs 5 ounces (140 grams).

First American Edition, 1998
Published in the United States by Dorling Kindersley Publishing, Inc.
95 Madison Avenue, New York, NY 10016

APPENDIX D

MINOR LEAGUE TEAMS

Aberdeen IronBirds
Akron Aeros
Albuquerque Isotopes
Altoona Curve
Arkansas Travelers
Asheville Tourists
Auburn Doubledays
Augusta GreenJackets
Bakersfield Blaze
Batavia Muckdogs
Beloit Snappers
Billings Mustangs
Binghamton Mets
Birmingham Barons
Bluefield Orioles
Boise Hawks
Bowie Baysox
Brevard County Manatees
Bristol White Sox
Brooklyn Cyclones
Buffalo Bisons
Burlington Bees
Burlington Indians
Carolina Mudcats
Casper Rockies
Cedar Rapids Kernels
Charleston RiverDogs
Charlotte Knights
Chattanooga Lookouts
Clearwater Threshers
Clinton LumberKings
Colorado Springs Sky Sox
Columbus Catfish
Columbus Clippers
Connecticut Defenders
Corpus Christi Hooks
Danville Braves
Dayton Dragons
Daytona Cubs
Delmarva Shorebirds
Dunedin Blue Jays
Durham Bulls
Elizabethton Twins
Erie SeaWolves
Eugene Emeralds
Everett AquaSox
Fort Myers Miracle
Fort Wayne Wizards
Frederick Keys
Fresno Grizzlies
Frisco RoughRiders
Great Falls White Sox
Greeneville Astros
Greensboro Grasshoppers
Greenville Bombers
Hagerstown Suns
Harrisburg Senators
Helena Brewers
Hickory Crawdads
High Desert Mavericks
Hudson Valley Renegades
Huntsville Stars
Idaho Falls Chukars
Indianapolis Indians
Inland Empire 66ers
Iowa Cubs
Jacksonville Suns
Jamestown Jammers
Johnson City Cardinals
Jupiter Hammerheads
Kane County Cougars
Kannapolis Intimidators
Kingsport Mets
Kinston Indians
Lake County Captains
Lake Elsinore Storm
Lakeland Tigers
Lakewood BlueClaws
Lancaster JetHawks
Lansing Lugnuts
Las Vegas 51s
Lexington Legends
Louisville Bats
Lowell Spinners
Lynchburg Hillcats
Mahoning Valley Scrappers
Memphis Redbirds
Midland RockHounds
Mississippi Braves
Missoula Osprey
Mobile BayBears
Modesto A's
Montgomery Biscuits
Myrtle Beach Pelicans
Nashville Sounds
New Britain Rock Cats
New Hampshire Fisher Cats
New Jersey Cardinals
New Orleans Zephyrs
Norfolk Tides
Ogden Raptors
Oklahoma RedHawks
Omaha Royals
Oneonta Tigers
Orem Owls
Ottawa Lynx
Palm Beach Cardinals
Pawtucket Red Sox
Peoria Chiefs
Portland Beavers
Portland Sea Dogs
Potomac Nationals
Princeton Devil Rays
Pulaski Blue Jays
Rancho Cucamonga Quakes
Reading Phillies
Richmond Braves
Rochester Red Wings
Rome Braves
Round Rock Express
Sacramento River Cats
Salem Avalanche
Salem-Keizer Volcanoes
Salt Lake Stingers
San Antonio Missions
San Jose Giants
Sarasota Red Sox
Savannah Sand Gnats
Scranton/Wilkes-Barre Red Barons
South Bend Silver Hawks
Southwest Michigan Devil Rays
Spokane Indians
Springfield Cardinals
St. Lucie Mets
Staten Island Yankees
Stockton Ports
Swing of the Quad Cities
Syracuse SkyChiefs
Tacoma Rainiers
Tampa Yankees
Tennessee Smokies
Toledo Mud Hens
Trenton Thunder
Tri-City Dust Devils
Tri-City ValleyCats
Tucson Sidewinders
Tulsa Drillers
Vancouver Canadians
Vermont Expos
Vero Beach Dodgers
Visalia Oaks
West Michigan Whitecaps
West Virginia Power
West Tenn Diamond Jaxx
Wichita Wranglers
Williamsport Crosscutters
Wilmington Blue Rocks
Winston-Salem Warthogs
Wisconsin Timber Rattlers
Yakima Bears

APPENDIX E

2006 MAJOR LEAGUE BASEBALL TEAMS

Seattle Mariners

Safeco Field
P.O. Box 4100
Seattle, WA 98104
Phone: (206) 346-4000
seattlemariners.com

Tampa Bay Rays

Tropicana Field
One Tropicana Drive
St. Petersburg, FL 33705
Phone: (727) 825-3137
www.raysbaseball.com

Texas Rangers
Rangers Ballpark in Arlington
1000 Ballpark Way
Arlington, TX 76011
Phone: (817) 273-5222
texasrangers.com

Toronto Blue Jays
Rogers Centre
1 Blue Jays Way, Suite 3200
Toronto, Ontario, Canada M5
Phone: (416) 341-1000
bluejays.com

Baltimore Orioles

Oriole Park at Camden Yards
333 West Camden Street
Baltimore, MD 21201
Phone: (410) 685-9800
orioles.com

Boston Red Sox

Fenway Park
4 Yawkey Way
Boston, MA 02215
Phone: (617) 267-9440
redsox.com

Chicago White Sox

U.S. Cellular Field
333 West 35th Street
Chicago, IL 60616
Phone: (312) 674-1000
whitesox.com
loswhitesox.com

Cleveland Indians

Jacobs Field
2401 Ontario Street
Cleveland, OH 44115
Phone: (216) 420-4200
indians.com

Detroit Tigers

Comerica Park
2100 Woodward Avenue
Detroit, MI 48201
Phone: (313) 471-2000
tigers.com

Kansas City Royals

Kauffman Stadium
One Royal Way
Kansas City, MO 64129
Phone: (816) 921-8000
royals.com

Los Angeles Angels of Anaheim

Angel Stadium of Anaheim
2000 Gene Autry Way
Anaheim, CA 92806
Phone: (714) 940-2000
angelsbaseball.com

Minnesota Twins

Metrodome
34 Kirby Puckett Place
Minneapolis, MN 55415
Phone: (612) 375-1366
twinsbaseball.com

New York Yankees

Yankee Stadium
161st Street and River Avenue
Bronx, NY 10451
Phone: (718) 293-4300
yankees.com
yankeesbeisbol.com

Oakland Athletics

McAfee Coliseum
7000 Coliseum Way
Oakland, CA. 94621

Arizona Diamondbacks

Chase Field
401 East Jefferson Street
Phoenix, AZ 85001
Phone: (602) 462-6500
dbacks.com
losdbacks.com

Atlanta Braves

Atlanta Braves
Turner Field
755 Hank Aaron Drive
Atlanta, GA 30315
Phone: (404) 522-7630
braves.com

Chicago Cubs

Wrigley Field
1060 West Addison
Chicago, IL 60613-4397
Phone: (773) 404-2827
cubs.com
loscubs.com

Cincinnati Reds

Great American Ball Park
100 Main Street
Cincinnati, OH 45202-4109
Phone: (513) 765-7000
reds.com

Colorado Rockies

Coors Field
2001 Blake Street
Denver, CO 80205-2000
Phone: (303) 292-0200
coloradorockies.com

Florida Marlins

Dolphin Stadium
2269 Dan Marino Boulevard
Miami, FL 33056
Phone: (305) 626-7400
floridamarlins.com
marlinsbeisbol.com

Houston Astros

Minute Maid Park
501 Crawford Street
Houston, TX 77002
Phone: (713) 259-8000
astros.com
astrosdehouston.com

Los Angeles Dodgers

Dodger Stadium
1000 Elysian Park Avenue
Los Angeles, CA 90012-1199
Phone: (323) 224-1500
dodgers.com
losdodgers.com

Milwaukee Brewers

Miller Park
One Brewers Way
Milwaukee, WI 53214
Phone: (414) 902-4400
brewers.com

New York Mets

Shea Stadium
123-01 Roosevelt Avenue
Flushing, NY 11368-1699
Phone: (718) 507-6387
mets.com
losmets.com

Philadelphia Phillies

Citizens Bank Park
One Citizens Bank Way
Philadelphia, PA 19148
Phone: (215) 463-6000
phillies.com

Pittsburgh Pirates

PNC Park
115 Federal Street
Pittsburgh, PA 15212
Phone: (412) 323-5000
pirates.com

San Diego Padres

PETCO Park
100 Park Boulevard
San Diego, CA 92101
Phone: (619) 795-5000
padres.com
padresbeisbol.com

San Francisco Giants

AT&T Park
24 Willie Mays Plaza
San Francisco, CA 9410
Phone: (415) 972-2000
sfgiants.com
sfgigantes.com

St. Louis Cardinals

Busch Stadium
700 Clark Street
St. Louis, MO 63102
Phone: (314) 345-9600
stlcardinals.com

Washington Nationals

RFK Stadium
2400 East Capitol Street
Washington, D.C. 20003
Phone: (202) 349-0400
nationals.com

APPENDIX F

FINAL STANDINGS 2007

American League

East	W	L	PCT	GB	E#	L10	STRK	HOME	ROAD	X W-L
Boston	96	66	.593	-	-	6-4	L1	51-30	45-36	101-61
New York	94	68	.580	2.0	E	6-4	W2	52-29	42-39	97-65
Toronto	83	79	.512	13.0	E	6-4	L1	49-32	34-47	86-76
Baltimore	69	93	.426	27.0	E	4-6	L2	35-46	34-47	71-91
Tampa Bay	66	96	.407	30.0	E	3-7	W1	37-44	29-52	67-95

Central	W	L	PCT	GB	E#	L10	STRK	HOME	ROAD	X W-L
Cleveland	96	66	.593	-	-	6-4	W1	52-29	44-37	91-71
Detroit	88	74	.543	8.0	E	5-5	W1	45-36	43-38	89-73
Minnesota	79	83	.488	17.0	E	4-6	W1	41-40	38-43	80-82
Chicago	72	90	.444	24.0	E	6-4	L1	38-43	34-47	67-95
Kansas City	69	93	.426	27.0	E	3-7	L1	35-46	34-47	74-88

West	W	L	PCT	GB	E#	L10	STRK	HOME	ROAD	X W-L
Los Angeles	94	68	.580	-	-	4-6	L1	54-27	40-41	90-72
Seattle	88	74	.543	6.0	E	7-3	W5	49-32	39-42	79-83
Oakland	76	86	.469	18.0	E	2-8	W1	40-41	36-45	79-83
Texas	75	87	.463	19.0	E	5-5	L3	47-34	28-53	79-83

National League

East	W	L	PCT	GB	E#	L10	STRK	HOME	ROAD	X W-L
Philadelphia	89	73	.549	-	-	7-3	W1	47-34	42-39	87-75
New York	88	74	.543	1.0	E	4-6	L1	41-40	47-34	86-76
Atlanta	84	78	.519	5.0	E	5-5	L2	44-37	40-41	88-74
Washington	73	89	.451	16.0	E	5-5	L1	40-41	33-48	70-92
Florida	71	91	.438	18.0	E	6-4	W1	36-45	35-46	72-90

Central	W	L	PCT	GB	E#	L10	STRK	HOME	ROAD	X W-L
Chicago	85	77	.525	-	-	6-4	L1	44-37	41-40	87-75
Milwaukee	83	79	.512	2.0	E	5-5	W2	51-30	32-49	83-79
St. Louis	78	84	.481	7.0	E	7-3	W5	43-38	35-46	71-91
Houston	73	89	.451	12.0	E	7-3	W2	42-39	31-50	72-90
Cincinnati	72	90	.444	13.0	E	3-7	W1	39-42	33-48	75-87
Pittsburgh	68	94	.420	17.0	E	2-8	L4	37-44	31-50	70-92

West	W	L	PCT	GB	E#	L10	STRK	HOME	ROAD	X W-L
Arizona	90	72	.556	-	-	5-5	L2	50-31	40-41	79-83
Colorado	90	73	.552	0.5	E	9-1	W3	51-31	39-42	91-72
San Diego	89	74	.546	1.5	E	4-6	L3	47-34	42-40	89-74
Los Angeles	82	80	.506	8.0	E	3-7	L1	43-38	39-42	82-80
San Francisco	71	91	.438	19.0	E	4-6	W1	39-42	32-49	77-85

APPENDIX G

ATTENDANCE

American League

Team	Home	Road
Baltimore	2,153,139	2,537,225
Boston	2,930,588	2,923,997
Chicago	2,957,414	2,453,307
Cleveland	1,997,995	2,447,511
Detroit	2,595,937	2,365,374
Kansas City	1,372,638	2,538,644
LA Angels	3,406,790	2,242,532
Minnesota	2,285,018	2,267,564
New York	4,248,067	3,080,265
Oakland	1,976,625	2,575,051
Seattle	2,481,165	2,320,325
Tampa Bay	1,368,950	2,318,803
Texas	2,388,757	2,391,333
Toronto	2,302,212	2,313,400

National League

Team	Home	Road
Arizona	2,091,685	2,514,961
Atlanta	2,550,524	2,412,534
Chicago	3,123,215	2,739,251
Cincinnati	2,134,607	2,747,075
Colorado	2,104,362	2,626,037
Florida	1,164,134	2,507,861
Houston	3,022,763	2,550,582
Los Angeles	3,758,545	2,586,375
Milwaukee	2,335,643	2,418,581
New York	3,379,535	2,725,074
Philadelphia	2,701,815	2,447,717
Pittsburgh	1,861,549	2,579,952
St. Louis	3,407,104	2,738,560
San Diego	2,659,757	2,641,722
San Francisco	3,130,313	2,628,251
Washington	2,153,056	2,404,038

APPENDIX H

TAMPA BAY RAYS

TAMPA BAY RAYS 2006

Compare	Seat Location	Price
	Section: 119 Row: X LOWER BOX	**$50 ea.**
	Section: 121 Row: R LOWER BOX	**$55 ea.**
	Section: 119 Row: R LOWER BOX	**$55 ea.**
	Section: 113 Row: U LOWER BOX	**$59 ea.**
	Section: 113 Row: W LOWER BOX	**$59 ea.**
	Section: 117 Row: X LOWER BOX	**$59 ea.**
	Section: 115 Row: X LOWER BOX	**$59 ea.**
	Section: 115 Row: S LOWER BOX	**$64 ea.**
	Section: 117 Row: S LOWER BOX	**$64 ea.**
	Section: 102 Row: FF	$77 ea.
	Section: 102 Row: DD	$77 ea.
	Section: 102 Row: GG	$77 ea.

APPENDIX I

TAMPA BAY RAYS TEAM AND SEASON RECORDS

Team History:
1998-Present - Tampa Bay Devil Rays (AL)
Team Address: Tropicana Field
One Tropicana Drive
St. Petersburg, FL 33705
727-825-3137
Opened: March 3, 1990 (as the Florida Suncoast Dome)
Renamed: Tropicana Field (October 4, 1996)
First Official Game: March 31, 1998, 11-6 loss to the Detroit Tigers
Stadium Capacity: 45,200
Surface: Field Turf
Dimensions:
* Home to Left field - 315 Feet
* Home to Left Center Field - 370 Feet
* Home to Left Center Field - 415 Feet
* Home to Center Field - 407 Feet
* Home to Right Center Field - 409 Feet
* Home to Right Center Field - 370 Feet
* Home to Right Field - 322 Feet

Managing General Partner: Stuart Sternberg
Team President: Matt Silverman
Chief of Baseball Operations: Andrew Friedman
Senior Vice President of Baseball Operations: Gerry Hunsicker
Manager: Joe Maddon
Manager History:

Term	Manager	Reg.Season	Playoff
1998-2001	Larry Rothschild	205-294	
	(fired after 14 games in 2001 season)		
2001-2002	Hal McRae	113-196	
2002-2005	Lou Pinella	200-285	
2005-present	Joe Maddon	61-101	

Retired Numbers: (1)
42 - Jackie Robinson
Devil Rays in the Hall of Fame: (1)
2005 - Wade Boggs
World Series Titles: None
National League Pennants: None
Division Titles: None
Wild Card: None

		Reg. Season						Playoffs		
Year	Location	W	L	Pct.	GB	Pos.	Div.	W	L	Pct.
1998	Tampa Bay	63	99	.389	51	(5)	East			
1999	Tampa Bay	69	93	.426	29	(5)	East			
2000	Tampa Bay	69	92	.429	18	(5)	East			
2001	Tampa Bay	62	100	.383	34	(5)	East			
2002	Tampa Bay	55	106	.342	48	(5)	East			
2003	Tampa Bay	63	99	.389	38	(5)	East			
2004	Tampa Bay	70	91	.435	30.5	(4)	East			
2005	Tampa Bay	67	95	.414	28	(5)	East			
2006	Tampa Bay	61	101	.377	36	(5)	East			

* - Strike Shortened Season

APPENDIX J

TAMPA BAY RAYS INDIVIDUAL STATISTICS

Batters	BA	SLG	G	AB	R	H	TB	2B	3B	HR	RBI	BB	SO	SB	CS	E
J.Shields	.375	.375	21	8	2	3	3	0	0	0	0	0	3	0	0	0
D.Young	.317	.476	30	126	16	40	60	9	1	3	10	1	24	2	2	1
C.Crawford	.305	.482	151	600	89	183	289	20	16	18	77	37	85	58	9	3
R.Baldelli	.302	.533	92	364	59	110	194	24	6	16	57	14	70	10	1	5
G.Norton	.296	.520	98	294	47	87	153	15	0	17	45	35	69	1	5	3
T.Wigginton	.275	.498	122	444	55	122	221	25	1	24	79	32	97	4	3	8
J.Paul	.260	.342	58	146	15	38	50	9	0	1	8	14	39	1	2	0
J.Cantu	.249	.404	107	413	40	103	167	18	2	14	62	26	91	1	1	13
B.Upton	.246	.291	50	175	20	43	51	5	0	1	10	13	40	11	3	13
D.Navarro	.244	.342	56	193	23	47	66	7	0	4	20	20	33	1	1	7
D.Hollins	.228	.423	121	333	37	76	141	20	0	15	33	19	64	3	3	4
T.Lee	.224	.364	114	343	35	77	125	11	2	11	31	42	73	5	2	2
B.Zobrist	.224	.311	52	183	10	41	57	6	2	2	18	10	26	2	3	9
J.Gomes	.216	.431	117	385	53	83	166	21	1	20	59	61	116	1	5	0
T.Perez	.212	.286	99	241	31	51	69	12	0	2	16	5	44	1	0	8
S.Burroughs	.190	.238	8	21	3	4	5	1	0	0	1	4	7	1	0	1
S.Riggans	.172	.207	10	29	3	5	6	1	0	0	1	4	7	0	0	0
K.Witt	.148	.279	19	61	5	9	17	2	0	2	5	0	21	0	0	2
D.Miceli	---	.000	33	0	0	0	0	0	0	0	0	0	0	0	0	0
B.Meadows	---	.000	53	0	0	0	0	0	0	0	0	0	0	0	0	0
S.Dunn	---	.000	7	0	0	0	0	0	0	0	0	0	0	0	0	0
J.Hammel	---	.000	9	0	0	0	0	0	0	0	0	0	0	0	0	0
D.Waechter	---	.000	11	0	0	0	0	0	0	0	0	0	0	0	0	0
E.Jackson	---	.000	23	0	0	0	0	0	0	0	0	0	0	0	0	0
J.Switzer	---	.000	40	0	0	0	0	0	0	0	0	0	0	0	0	0
C.Harville	---	.000	32	0	0	0	0	0	0	0	0	0	0	0	0	0
J.Seo	.000	.000	17	1	0	0	0	0	0	0	0	0	0	0	0	0
T.Walker	---	.000	20	0	0	0	0	0	0	0	0	0	0	0	0	0
S.Camp	---	.000	75	0	0	0	0	0	0	0	0	0	0	0	0	0
S.McClung	.000	.000	39	1	0	0	0	0	0	0	0	0	0	0	0	1
B.Stokes	---	.000	5	0	0	0	0	0	0	0	0	0	0	0	0	0
L.Ordaz	.000	.000	1	2	0	0	0	0	0	0	0	0	0	0	0	0
J.Colome	---	.000	1	0	0	0	0	0	0	0	0	0	0	0	0	0
T.Harper	---	.000	30	0	0	0	0	0	0	0	0	0	0	0	0	0
T.Corcoran	.000	.000	21	4	0	0	0	0	0	0	0	0	3	0	0	0
J.Childers	---	.000	5	0	0	0	0	0	0	0	0	0	0	0	0	0
C.Fossum	.000	.000	25	2	0	0	0	0	0	0	0	0	0	0	0	0
R.Lugo	---	.000	64	0	0	0	0	0	0	0	0	0	0	0	0	0
J.Salas	---	.000	8	0	0	0	0	0	0	0	0	0	0	0	0	0
S.Kazmir	.000	.000	24	3	0	0	0	0	0	0	0	0	2	0	0	0
C.Orvella	---	.000	22	0	0	0	0	0	0	0	0	0	0	0	0	0
J.Howell	---	.000	8	0	0	0	0	0	0	0	0	0	0	0	0	0

APPENDIX K

BASEBALL BAT SIZES

Boys
Batter's Height

Weight in Pounds	3'5"-3'8"	3'9"-4'	4'1"-4'4"	4'5"-4'8"	4'9"-5'	5'1"-5'4"	5'5"-5'8"	5'9"-6'
under 60	27"	28"	29"	29"				
61-70	27"	28"	29"	29"	30"			
71-80	28"	28"	29"	30"	30"	31"		
81-90	28"	29"	29"	30"	30"	31"	32"	
91-100	28"	29"	30"	30"	31"	31"	32"	
101-110	29"	29"	30"	30"	31"	31"	32"	
111-120	29"	29"	30"	30"	31"	31"	32"	
121-130	29"	30"	30"	30"	31"	32"	32"	
131-140	29"	30"	30"	31"	31"	32"	33"	33"
141-150		30"	30"	31"	31"	32"	33"	34"
151-160		30"	31"	31"	32"	32"	33"	34"
over 160			31"	31"	32"	32"	33"	34"

Girls
Batter's Height

Weight in Pounds	3'10"-4'	4'1"-4'4"	4'5"-4'8"	4'9"-5'	5'1"-5'4"	5'5"-5'9"
under 40	26"	27"	28"			
40-45	27"	28"	29"	30"		
46-50	27"	28"	29"	30"		
51-60	27"	28"	29"	30"	31"	
61-70	28"	29"	30"	31"	32"	
71-80	28"	29"	30"	31"	32"	33"
81-90	29"	30"	31"	32"	33"	33"
91-100	29"	30"	31"	32"	33"	34"
101-110		30"	31"	32"	33"	34"
111-120		31"	32"	33"	34"	34"
121-130		31"	32"	33"	34"	34"

APPENDIX R

APPENDIX L

MAJOR LEAGUE BASEBALL ALL-TIME RECORDS

Major League Baseball All-time Statistics
(complete through 2006 regular season)

From The Sports Network

Batting

Home Runs		Runs Batted In	
--------------		------------------	
Hank Aaron	755	Hank Aaron	2297
Barry Bonds	734	Babe Ruth	2213
Babe Ruth	714	Cap Anson	2076
Willie Mays	660	Lou Gehrig	1995
Sammy Sosa	588	Stan Musial	1951
Frank Robinson	586	Ty Cobb	1938
Mark McGwire	583	Barry Bonds	1930
Harmon Killebrew	573	Jimmie Foxx	1922
Rafael Palmeiro	569	Eddie Murray	1917
Ken Griffey, Jr.	563	Willie Mays	1903
Reggie Jackson	563	Mel Ott	1860
Mike Schmidt	548	Carl Yastrzemski	1844
Mickey Mantle	536	Ted Williams	1839
Jimmie Foxx	534	Rafael Palmeiro	1835
Willie McCovey	521	Dave Winfield	1833
Ted Williams	521	Al Simmons	1827
Ernie Banks	512	Frank Robinson	1812
Eddie Mathews	512	Honus Wagner	1732
Mel Ott	511	Reggie Jackson	1702
Eddie Murray	504	Cal Ripken, Jr.	1695
Lou Gehrig	493	Tony Perez	1652

APPENDIX M

"WHO'S ON FIRST"

By Abbot and Costello

Abbot: Well Costello, I'm going to New York with you. The Yankee's manager gave me a job as coach for as long as you're on the team.
Costello: Look Abbott, if you're the coach, you must know all the players.
Abbot: I certainly do.
Costello: Well you know I've never met the guys. So you'll have to tell me their names, and then I'll know who's playing on the team.
Abbot: Oh, I'll tell you their names, but you know strange as it may seem, they give these ball players now-a-days very peculiar names.
Costello: You mean funny names?
Abbot: Strange names, pet names ... like Dizzy Dean ...
Costello: His brother Daffy
Abbot: Daffy Dean ...
Costello: And their French cousin.
Abbot: French?
Costello: Goofe'
Abbot: Goofe' Dean. Well, let's see, we have on the bags, Who's on first, What's on second, I Don't Know is on third ...
Costello: That's what I want to find out.
Abbot: I say Who's on first, What's on second, I Don't Know's on third.
Costello: Are you the manager?
Abbot: Yes.
Costello: You gonna be the coach too?
Abbot: Yes.
Costello: And you don't know the fellows' names.
Abbot: Well I should.
Costello: Well then who's on first?
Abbot: Yes.
Costello: I mean the fellow's name.
Abbot: Who.
Costello: The guy on first.
Abbot: Who.
Costello: The first baseman.
Abbot: Who.
Costello: The guy playing ...
Abbot: Who is on first!
Costello: I'm asking you who's on first.
Abbot: That's the man's name.
Costello: That's who's name?
Abbot: Yes.
Costello: Well go ahead and tell me.
Abbot: That's it.
Costello: That's who?
Abbot: Yes.
PAUSE
Costello: Look, you got afirst baseman?
Abbot: Certainly.
Costello: Who's playing first?
Abbot: That's right.
Costello: When you pay off the first baseman every month, who gets the money?
Abbot: Every dollar of it.
Costello: All I'm trying to find out is the fellow's name on first base.
Abbot: Who.
Costello: The guy that gets the money...
Abbot: That's it.
Costello: Who gets the money ...
Abbot: He does, every dollar of it. Sometimes his wife comes down and collects it.
Costello: Who's wife?
Abbot: Yes.
PAUSE
Abbot: What's wrong with that?
Costello: Look, all I wanna know is when you sign up the first baseman, how does he sign his name to the contract?
Abbot: Who.
Costello: The guy.
Abbot: Who.
Costello: How does he sign ...
Abbot: That's how he signs it.
Costello: Who?
Abbot: Yes.
PAUSE
Costello: All I'm trying to find out is what's the guy's name on first base.
Abbot: No, What is on second base.
Costello: I'm not asking you who's on second.
Abbot: Who's on first.
Costello: One base at a time!
Abbot: Well, don't change the players around.
Costello: I'm not changing nobody!
Abbot: Take it easy, buddy.
Costello: I'm only asking you, who's the guy on first base?
Abbot: That's right.
Costello: Ok.
Abbot: Alright.
PAUSE
Costello: What's the guy's name on first base?
Abbot: No. What is on second.
Costello: I'm not asking you who's on second.
Abbot: Who's on first.
Costello: I don't know.
Abbot: Oh, he's on third, we're not talking about him.
Costello: Now how did I get on third base?
Abbot: Why, you mentioned his name.
Costello: If I mentioned the third baseman's name, who did I say is playing third?
Abbot: No, Who's playing first.
Costello: What's on first?
Abbot: What's on second.
Costello: I don't know.
Abbot: He's on third.
Costello: There I go, back on third again!
PAUSE
Costello: Would you just stay on third base and don't go off it?!

Abbot: Alright, what do you want to know?
Costello: Now who's playing third base?
Abbot: Why do you insist on putting Who on third base?
Costello: What am I putting on third?
Abbot: No, What is on second.
Costello: You don't want who on second?
Abbot: Who is on first.
Costello: I don't know.
Together: Third base!
PAUSE
Costello: Look, you got an outfield?
Abbot: Sure.
Costello: The left fielder's name?
Abbot: Why.
Costello: I just thought I'd ask you.
Abbot: Well, I just thought I'd tell ya.
Costello: Then tell me who's playing left field.
Abbot: Who's playing first.
Costello: I'm not --stay out of the infield! I want to know what's the guy's name in left field?
Abbot: No, What is on second.
Costello: I'm not asking you who's on second.
Abbot: Who's on first!
Costello: I don't know.
Together: Third base!
PAUSE
Costello: And the left fielder's name?
Abbot: Why.
Costello: Because!
Abbot: Oh, he's center field.
PAUSE
Costello: Look, look, look. You gotta pitcher on the team?
Abbot: Sure.
Costello: The pitcher's name?
Abbot: Tomorrow.
Costello: You don't want to tell me today?
Abbot: I'm telling you now.
Costello: Then go ahead.
Abbot: Tomorrow!
Costello: What time?
Abbot: What time what?
Costello: What time tomorrow are you gonna tell me who's pitching?
Abbot: Now listen. Who is not pitching, Who....
Costello: I'll break your arm if you say who's on first! I want to know what's the pitcher's name?
Abbot: What's on second.
Costello: I don't know.
Together: Third base!
PAUSE
Costello: Gotta a catcher?
Abbot: Certainly.
Costello: The catcher's name?
Abbot: Today.
Costello: Today, and tomorrow's pitching.
Abbot: Now you've got it.
Costello: All we got is a couple of days on the team.
PAUSE
Costello: You know I'm a catcher too.
Abbot: So they tell me.
Costello: I get behind the plate to do some fancy catching, Tomorrow's pitching on my team and a heavy hitter gets up. Now the heavy hitter bunts the ball. When he bunts the ball, me, being a good catcher, I'm gonna throw the guy out at first. So I pick up the ball and throw it to who?
Abbot: Now that's the first thing you've said right.
Costello: I don't even know what I'm talking about!
PAUSE
Abbot: That's all you have to do.
Costello: Is to throw the ball to first base.
Abbot: Yes!
Costello: Now who's got it?
Abbot: Naturally.
PAUSE
Costello: Look, if I throw the ball to first base, somebody's gotta get it. Now who has it?
Abbot: Naturally.
Costello: Who?
Abbot: Naturally.
Costello: Naturally?
Abbot: Naturally.
Costello: So I pick up the ball and I throw it to Naturally.
Abbot: No you don't you throw the ball to Who.
Costello: Naturally.
Abbot: That's different.
Costello: That's what I said.
Abbot: Your not saying it right ...
Costello: I throw the ball to Naturally.
Abbot: You throw it to Who.
Costello: Naturally.
Abbot: That's it.
Costello: That's what I said!
Abbot: Listen, you ask me.
Costello: I throw the ball to who?
Abbot: Naturally.
Costello: Now you ask me.
Abbot: You throw the ball to Who?
Costello: Naturally.
Abbot: That's it.
Costello: Same as you! Same as you! I throw the ball to who. Whoever it is drops the ball and the guy runs to second. Who picks up the ball and throws it to What. What throws it to I Don't Know. I Don't Know throws it back to Tomorrow, Triple play. Another guy gets up and hits a long fly ball to Because. Why? I don't know! He's on third and I DON'T GIVE A DARN!
Abbot: What?
Costello: I said I don't give a darn!
Abbot: Oh, that's our shortstop.

APPENDIX N

CASEY AT THE BAT

Ernest Lawrence Thayer

The outlook wasn't brilliant for the Mudville nine that day;
The score stood four to two, with but one inning more to play;
And so, when Cooney died at first, and Burrows did the same,
A sickly silence fell upon the patrons of the game.

A straggling few got up to go in deep despair. The rest
Clung to the hope which springs eternal in the human breast;
They thought, if only Casey could but get a whack, at that,
They'd put up even money now, with Casey at the bat.

But Flynn preceded Casey, as did also Jimmy Blake,
And the former was a pudding, and the latter was a fake;
So upon that stricken multitude grim melancholy sat,
For there seemed but little chance of Casey's getting to that bat.

But Flynn let drive a single, to the wonderment of all
and Blake, the much despised, tore the cover off the ball;
And when the dust had lifted, and they saw what had occurred,
There was Jimmy safe on second, and Flynn a-hugging third.

Then from the gladdened multitude went up a joyous yell;
It bounded from the mountain top, and rattled in the dell;
It struck upon the hillside, and recoiled upon the flat;
For Casey, mighty Casey, was advancing to the bat.

There was ease in Casey's manner as he stepped into his place;
There was pride in Casey's bearing, and a smile on Casey's face;
And when, responding to the cheers, he lightly doffed his hat,
No stranger in the crowd could doubt 'twas Casey at the bat.

Ten thousand eyes were on him as he rubbed his hands with dirt,
Five thousand tongues applauded when he wiped them on his shirt;
Then while the writhing pitcher group the ball into his hip,
Defiance gleamed in Casey's eye, a sneer curled Casey's lip.

And now the leather-covered sphere came hurtling through the air,
And Casey stood a-watching it in haughty grandeur there;
Close by the sturdy batsman the ball unheeded sped.
"That ain't my style," said Casey. "Strike One!" the umpire said.

From the benches, black with people, there went up a muffled roar,
Like the beating of the storm waves on a stern and distant shore;
"Kill him! Kill the umpire!" shouted someone on the stand;
And it's likely they'd have killed him had not Casey raised his hand.

With a smile of Christian charity great Casey's visage shone;
He stilled the rising tumult; he bade the game go on;
He signaled to the pitcher, and once more the spheroid flew;
But Casey still ignored it, and the umpire said, "Strike Two!"

" Fraud!" cried the maddened thousands, and the echo answered, "Fraud!"
But a scornful look from Casey, and the audience was awed.
They saw his face grow stern and cold, they saw his muscled strain,
And they knew that Casey wouldn't let that ball go by again.

The sneer is gone from Casey's lips, his teeth are clenched in hate,
He pounds with cruel violence his bat upon the plate;
And now the pitcher holds the ball, and now he lets it go,
And now the air is shattered by the force of Casey's blow.

Oh! Somewhere in this favored land sun is shining bright;
The band is playing somewhere, and somewhere hearts are light;
And somewhere men are laughing, and somewhere children shout,
But there is no joy in Mudville - Mighty Casey has struck out!

APPENDIX O

BASEBALL AT 9-11-2001

Baseball proved to be theraputic

By Ken Gurnick - MLB.com - 9/18/2001 4:00 am ET

LOS ANGELES - Baseball did its part in the healing of a wounded nation Monday and went back to work.

Following the financial markets in heeding President Bush's call for America to "get on about its life," baseball games resumed. They provided the country a welcome diversion from the grief and mourning that followed last Tuesday's terrorist attacks. "I think a lot of us needed this. I know I needed this," said William Patterson of Quakertown, NJ, attending the Atlanta -Philadelphia game. "I just needed something positive to take my mind off things."

In six stadiums Monday, baseball provided something positive. "Despite heavy hearts, we've gotten out of the dirt, brushed ourselves off and we're hoping in some small way to inspire a nation to do the same thing," Dodgers Hall of Fame broadcaster Vin Scully announced in a video address before the Dodgers - Padres game. "It's a bittersweet evening, but the first step toward normalcy."

Even before the games started, players stood in stadium outfields elbow-to-elbow with local police and firefighters to unfurl American flags in emotional ceremonies honoring the memory of those who lost their lives last week and saluting emergency personnel who risked their lives to save them.

Fans - many draped in red, white and blue - waved small flags and chanted "USA...USA." The traditional singing of the national anthem was accompanied by "God Bless America." Stadium organists played patriotic medleys. Flags were stitched onto the backs of player uniforms and caps. A moment of silence was held in memory of the victims.

Generally, players were touched by the pregame ceremonies, but uneasy about returning to competition at a time of national crisis. "It was just real nice to see the fans and the flags out there, being patriotic and the seventh-inning stretch that we did," said Arizona pitcher Randy Johnson, who won his 19th game. "What you're seeing is the country coming together and there was no doubt that would happen. It's a very close-knit country and it's very tragic that something like this has to happen to see all of this." And when all that was done, the players took the field and played ball.

Ending a six-day shutdown, six National League games were played Monday and the American League, where all four postseason slots are virtually locked up, resumes Tuesday.

Emotions ran high in Philadelphia and Pittsburgh. Fans in Los Angeles were noisy and left typically early. In Colorado, players described the mood as subdued. The most meaningful game played Monday had nothing to do with a pennant race. The shaken New York Mets, their home game with the Pirates transferred to Pittsburgh because of the World Trade Center catastrophe, wore caps with familiar initials NYPD and FDNY, a salute to the city's police and fire departments.

The day before, Mets players visited victims in hospitals and manager Bobby Valentine stayed at Shea Stadium, an emergency staging area, until 3 a.m. loading supplies for transport to relief workers downtown.

Monday, back in uniform, the Mets beat the Pirates, 4-1, with a ninth-inning rally. The heart and soul of the team, a relief worker of a different kind named John Franco, picked up the win on his 41st birthday. "We were just trying to do our jobs as good as we could do it," valentine said. "That's what we get paid to do, and I think that's what our fans want us to do. We had good spirit in the dugout, and there was good focus on the mound."

Debate had been lively about the timing of the resumption f play, which was ordered by Commissioner Bud Selig.

"Mentally, I think we were all pretty unstable on what was right and what was wrong and where we needed to be", said Phillies third baseman Scott Rolen, who homered twice in the Phillies' win over Atlanta. "This was a good one. I'm glad it's behind us now and we can continue on. I'm glad the support was shown throughout the country. The whole game there was just a feeling on the ball field that I haven't had before. I can't explain it. Maybe tomorrow we'll go back to a little more normalcy."

Said Colorado manager Buddy Bell, after the Rockies' loss to Arizona, "As the night wore on, I think we all realized this was something we needed to do, we should be here."

Although games went on without a hitch, the players said these weren't normal games. Many described the early innings as "eerie." Dodgers manager Jim Tracy said there was "a mild sense of uncertainty."

"The pregame was touching, and everybody enjoyed being part of it," said Padres reliever Trevor Hoffman, who saved San Diego's 6-4 win over Los Angeles. "But, this didn't feel the same. Something is always in the back of your mind now. The flow of information has slowed down to the point where we're not glued to the TV for updates, so it's good if we can give the nation an outlet. But there's something there now, and it's hard to play the game the way it's meant to be played."

(continued on next page)

Taken from : www.mlb.com ("Our flag was still there..."September 2001)
http://indians.mlb.com/NASApp/mlb/mlb/official_info/mlb_official_info_relief.jsp

Mortality has not been a traditional topic of clubhouse conversation, but it is now, as ballplayers realize they can become targets and victims like anyone else. For the first time in their careers, many had to produce picture-identification to gain entry into their stadiums.

Security in ballparks was tightened throughout baseball. Backpacks, occasionally handed out as promotional items, now are banned. Bags are searched. In Los Angeles, police department bomb-sniffing dogs swept Dodger Stadium, including team clubhouses and the press box. "We've done everything we can without changing the fundamental nature of the event," said Sandy Alderson, Major League Baseball's executive vice president of baseball operations. "fans can have confidence that the ballpark has always been safe, and it's safer today than last week."

On the field, pennant races that seemed so important a week earlier, then were rendered virtually meaningless even to the players involved, nudged back into the picture Monday. The Arizona Diamondbacks strengthened their grip on the National League West with a 4-3 win in Colorado, extending their lead over idle San Francisco to two games. The six-day stoppage allowed Arizona manager Bob Brenly to rework his rotation so Randy Johnson, Monday night's winner, could start five of the team's last 19 games. "It's hard to see what's going on in the world and jump back into the pennant race," said Arizona outfielder Luis Gonzalez.

The Giants will pick up the season Tuesday night hosting Central Division leader Houston at pac Bell Park, with Barry Bonds resuming his chase of Mark McGwire's single-season home run record of 70. With 18 games remaining, Bonds has 63 home runs, including three in his most recent game and six in his last seven games. San Francisco is now tied for the Wild Card lead with St. Louis, which edged Milwaukee, 2-1. San Francisco and St. Louis are two games ahead of the Chicago Cubs, who play Cincinnati Tuesday in Chicago. The Dodgers, who must end the season with nine consecutive road games when last week's games are made up, lost a home game Monday night, wasting 11 strikeouts in six innings from Kevin Brown in a 6-4 loss to San Diego. Padres starting pitcher Jason Middlebrook allowed two hits over six innings in his winning Major League debut.

In the NL East, the Phillies cut into Atlanta's lead, now 2 1/2 games, with a 5-2 win over the Braves, as 15 game winner Robert Person out dueled 17-game winner Greg Maddux. "I've had a lot of games here and this was as emotional as I've ever been," said Phillies manager Larry Bowa. "I'm glad it's over and I'm glad we won. It was a tough game to get through."

Attendance at the games ranged from a sparse gathering of 3,013 at Olympic Stadium in Montreal to 40,676 at Dodger Stadium, where the crowd took the pennant race seriously enough to boo losing reliever Chan Ho Park when he left the game with an Achillies strain. Dodger fans weren't alone. When Atlanta's Chipper Jones homered in the first inning at Veterans Stadium, Philadelphia fans did what they're known for - they booed. "Yeah, they were coming back into their own," said Person, who gave up the homer.

Tony Gwynn, playing in Los Angeles, said he was surprised at how normal the game seemed to him. There were even a few beach ball sightings. "The fans got into it when the Dodgers rallied, they booed me in the on-deck circle, and they left in the seventh inning like they always (do)," Gwynn said. "It's almost like they didn't miss a beat. Like a typical regular season game. But now's the hard part. You leave the park, you turn on the radio and - bam - you're right back in it."

Taken from : www.mlb.com ("Our flag was still there..."September 2001)
http://indians.mlb.com/NASApp/mlb/mlb/official_info/mlb_official_info_relief.jsp

APPENDIX P

BASEBALL DEFINITIONS

Ahead of the Count - Said of the pitcher when there are more strikes than balls on the batter. Also, describing the batter when there are more balls than strikes.

Around the Horn - Term used to describe a double play in which a ground ball is fielded by the third baseman who throws to second base who throws to first base.

On Deck - Term given to the player that is schedule to be the next hitter.

Bang-Bang Play - A very close tag or force play when the runner and baseball arrive almost simultaneously.

Clean-Up Hitter - The player that bats fourth in the batting order. He is the player most likely to bat with players on base and have the opportunity to "clean or clear" the bases with a hit.

Fungo - A type of bat used to hit fly and ground balls, used particularly during batting practice.

Grand Slam - A home run that occurs with the bases loaded, producing four runs scored.

Horsehide - The ball itself. Baseballs are covered with horsehide or cowhide.

Knuckleball - A slowly thrown pitch that has little or no spin, causing it to wobble and dip unpredictably. It is gripped with the fingernails or knuckles.

National Pastime - A term commonly applied to baseball in the United States. It was first used in 1857.

Round-Tripper - A home run, from the fact that the batter leaves and returns home on the same "ticket".

Squeeze Play - Play where the batter attempts to score a runner from third base by bunting. The runner sprints for home with the pitch and the batter bunts the ball to a place where fielders can not throw out the runner.

Switch-hitter - A batter that hits both right and left-handed. A switch-hitter usually hits right handed against left-handed pitches and left-handed against right-handed pitchers.

Texas Leaguer - A poorly hit ball that loops meekly over the infield and lands for a hit.

Utility Player - A substitute that is a valuable member because of his/her ability to play several different positions.

APPENDIX Q

BASEBALL BOOKS

20,000 Baseball Cards under the Sea
by Jon Buller & Susan Schade

A Baseball All-Star
by Brendan January

Albert's Ballgame
by Leslie Tyron

Alex Rodriguez
by Jeffrey Zeuhlke

Arthur and the Seventh Inning Stretcher
by Stephen Krensky

Arthur Makes the Team
by Marc Brown

Babe Ruth and the Ice Cream Mess
by Dan Gutman

Ballpark
by Elisha Cooper

Ballpark: The story of America's Baseball Fields
by Lynn Curlee

Baseball
by James Kelley

Baseball
by Tom Owens

Baseball Ballerina
by Kathryn Cristaldi

Baseball Camp on the Planet of Eyeballs
by Jon Buller

Baseball for Everybody: Tom Glavine's Guide to America's Game
by Tom Glavine

Baseball's Best
by Andrew Gutelle

Baseball's boneheads, bad boys, & just plain crazy guys
by George Sullivan

Baseball's Greatest Hitters
by S.A. Kramer

Baseball's Greatest Pitchers
by S.A. Kramer

Bases Loaded: Great Baseball of the 20th Century
by Mel Cebulash

Bats About Baseball
by Jean Little & Claire Mackay

Black Diamond
by Patricia & Fred McKissack

Bob Feller
by Morris Eckhouse

Cal Ripken Jr.: My Story
with Dan Gutman

Cal Ripken Jr.: Play Ball
with Mike Bryan

Cam Jansen/Babe Ruth Baseball
by David Alder

Curious George Plays Baseball
edited by Margaret Rey & Allan Shalleck

Derek Jeter
by Michael Bradley

Diamond Life: baseball's sights, sounds and swings
by Charles R. Smith

Dog on Third Base
by Constance Hise

Extra Innings
by Robert Newton Peck

Frank and Ernest Play Ball
by Alexandra Day

Frank Thomas: Power Hitter
by Bill Gutman

Glory Days: the Akron Yankees
by Richard McBane

Ronald Morgan Goes to Bat
by Patricia Reilly Giff

Sammy Sosa
by Carrie Muskat

Satchel Paige
by Kathryn Long Humphrey

Shoeless Joe and Black Betsy
by Phil Bildner

Take Me Out to the Ballgame
by Maryann Kovalski

Take Me Out to the Ballgame
by Jack Norworth

Teammates
by Peter Goldenbook

That Sweet Diamond:baseball poems
by Paul B. Janeczko

The Baseball Birthday Party
by Annabella Prager

The Baseball Counting Book
by Barbara Barbieri McGrath

The Berenstein Bears Play Ball
by Stan & Jan Berenstein

The Boy Who Saved Baseball
by John Ritter

The Case of the Unnatural
by David D. Connell

The Dog that Pitched a No Hitter
by Matt Christopher

The Everything Kids' Baseball Book
by Richard Mintzer

The Field Beyond the Outfield
by George Sullivan

The Fireplug is First Base
by P.J. Peterson

The Fox Under First Base
by Jim Latimer

The Journal of Biddy Owens: the Negro Leagues
by Walter Dean Myers

The Jungle Baseball Game
by Tom Paxton

The Math Curse
by Jon Scieszka & Lane Smith

The Not-So-Minor Leagues
by Douglas Gay & Kathlyn Gay

The Rainy Day Grump
by Deborah Eaton

The Ripken Way: A Manual for Baseball and Life
by Cal Ripken Sr.

The Shot Heard 'Round the World
by Phil Bildner

The Spy on Third Base
by Matt Christopher

The World of Baseball
by James Buckley

Winners Take All
by Fred Bowen

World Series
by James Buckley

Young Cam Jansen and Baseball Mystery
by David Alder

Zachary's Ball
by Matt Tavares

APPENDIX R

BASEBALL-RELATED WEBSITES

Official Site of Major League Baseball
http://www.mlb.com

Official Site of the Akron Aeros
http://www.akronaeros.com

Official Site of Minor League Baseball
http://www.minorleaguebaseball.com

Official Site of the Cleveland Indians
http://www.clevelandindians.com

Exploratorium: Science of Baseball
http://www.exploratorium.edu/baseball
Scientific research and interactive experiments. Requires Shockwave for some of the features.

The Baseball Archive
http://baseball1.com
Sabermetrics is the mathematical and statistical analysis of baseball records.

General Essays & Background Theory
http://www.stathead.com/bbeng/general.htm

Baseball Think Factory
http://www.baseballthinkfactory.com
Dedicated to the thoughtful analysis of baseball, both real and imagined. Assistance for both sabermetricians and game players is provided here.

Stathead Consulting
http://www.stathead.com
Providing perfomance analysis and forecasting services to the professional industry, includes the Baseball Engineering Library.

Society of American Baseball Research (SABR)
http://sabr.org
The Society for American Baseball Research (SABR) was formed in August 1971 in Cooperstown, New York. It now consists of more than 6,700 members.

Supervision
http://www.questec.com/
A pitch tracking system that produces a real-time computer animation of the pitch immediately after the actual pitch is thrown.

Rules of the Game - Official Baseball Rules
http://mlb.mlb.com/NASApp/mlb/mlb/official_info/official_rules/foreword.jsp
Links for baseball lingo definitions as well as links for official major league rules.

3000 Hit Club
http://www.baseballhalloffame.org/exhibits/online_exhibits/3000_hit_club/index.htm
Baseball Hall of Fame page paying tribute to the exclusive 3000 hit club membership

National Baseball Hall of Fame
http://www.baseballhalloffame.org
Information regarding the Baseball Hall of Fame and the game of baseball itself.

Negro Leagues Baseball Museum
http://www.NLBM.com

Negro Baseball Leagues
http://www.blackbaseball.com

Official Baseball History: The National Baseball Hall of Fame and Museum
http://www.baseballhalloffame.org/history/index.htm
The official web site of the Hall of Fame. Don't miss the Members' Gallery (http://www.baseballhalloffame.org/hofers_and_honorees/plaques/index.htm)

John Skilton's Baseball Links
http://www.baseball-links.com/
The Web's most comprehensive collection of links to baseball resources.

Miscellaneous British Baseball Federation
http://www.bbf.org
The British Baseball Federation is the governing body for baseball in Great Britain.

Baseball Prospectus
http://www.baseballprospectus.com
Current season analysis.

An Internet activity about Baseball, Math and the San Francisco Giants
http://www.kn.pacbell.com/wired/baseball/

Box Scores, game logs, history and more
http://www.retrosheet.org

Players. History, stats and more
http://www.thebaseballpage.com

APPENDIX S

STADIUM SEATING CHART—TAMPA BAY RAYS

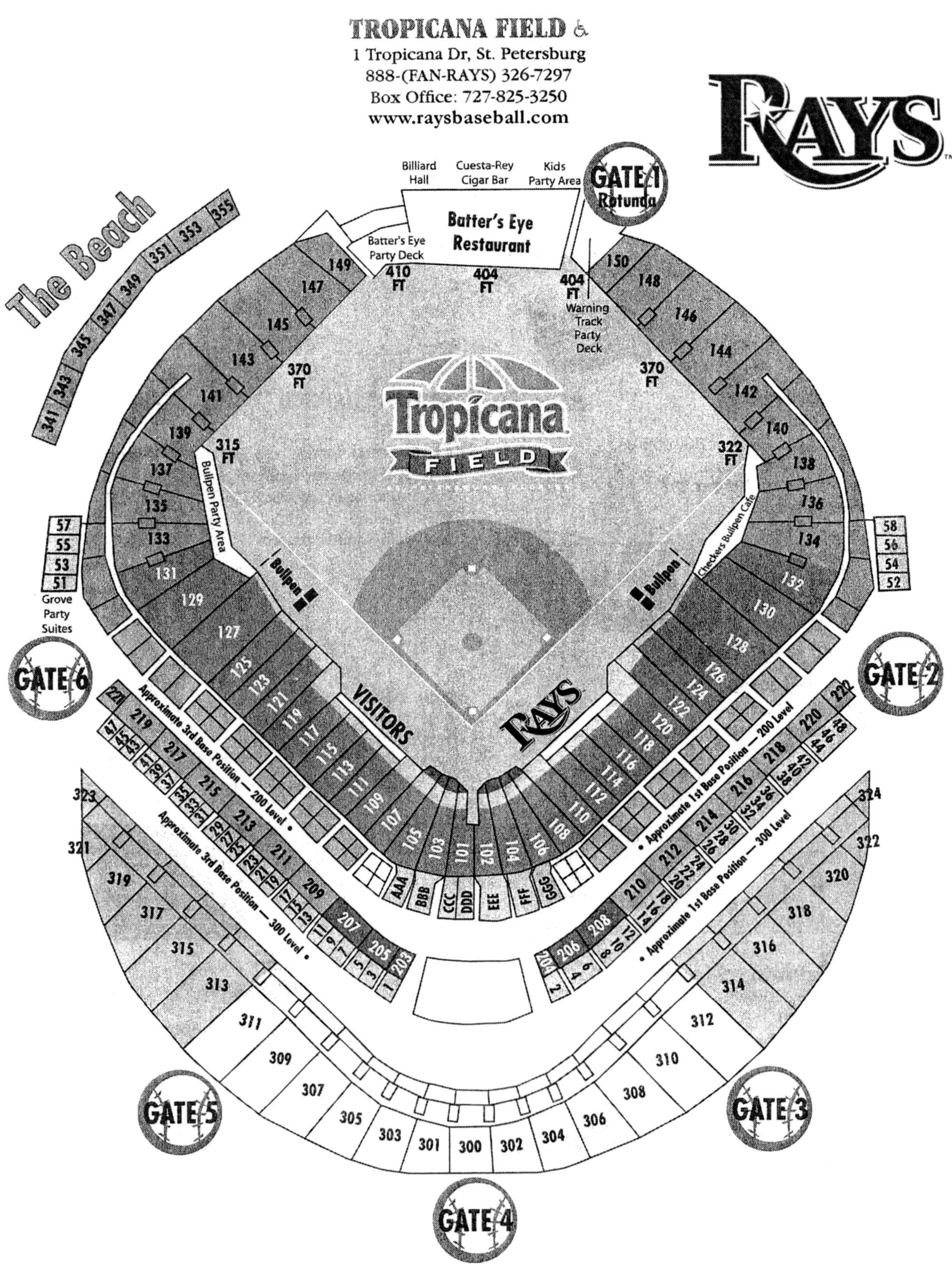

CPSIA information can be obtained at www.ICGtesting.com
Printed in the USA
LVOW090534211112

308281LV00002B/17/P

9 780982 744123